Naturalists' Handbooks 28

Studying invertebrates

C. PHILIP WHEATER

PENNY A. COOK

Pelagic Publishing

www.pelagicpublishing.com

Published by Pelagic Publishing
www.pelagicpublishing.com
PO Box 725, Exeter, EX1 9QU

Studying invertebrates
Naturalists' Handbooks 28

Series editors
S. A. Corbet and R. H. L. Disney

ISBN 978-1-78427-082-7

Digital reprint edition of ISBN 0-85546-313-9 (2003)

Illustrations by Jo Wright and others

British Library Cataloguing in Publication Data
A catalogue record for this book is available from the British Library.

Contents

Editors' preface

The books in this series are designed to encourage readers to undertake their own studies of natural history. Each one describes some relevant techniques, but they have not enough space to cover the substantial body of more generally applicable ideas and approaches that underlies the design and analysis of such field studies. By describing a selection of these general methods, *Studying invertebrates* aims to support those venturing into ecological fieldwork for the first time. The authors have plenty of experience in helping beginners to plan, carry out and interpret ecological surveys and experiments, and we hope this handbook will serve as a welcome companion and guide, especially for those who lack confidence in their knowledge of statistical and other methods.

S.A.C.
R.H.L.D.
April 2003

Dedication and acknowledgements

We dedicate this book to the late J. Gordon Blower who inspired generations of students and naturalists.

We would like to thank all those who have encouraged our interest in ecology, ecological techniques and statistical analyses over the years, especially Gordon Blower, Glyn Evans, Derek Yalden, Robin Baker, Mike Hounsome, Ian Harvey and Nina Wedell. We are also grateful to all those who contributed directly to this book either by reading and commenting on sections, or by providing advice and materials: James Bell, Mike Dobson, Hugh Jones, Liz Price, Helen Read, and especially Mark Langan. We particularly thank Jo Wright who produced most of the line drawings and the editors (Sally Corbet and Henry Disney) without whose careful advice and considerable assistance this text would have been less accessible and comprehensive.

Jo Wright produced figs.1–3, 6–12, 14–16, 19, 24–29, 34, 39, 41, 47, 50, 53, 55–58, 61–63, 67–68, 81, 93–101 and 103. We would like to thank all those who allowed us to use their drawings from other books in this series including: Sophie Allington (figs. 21, 22, 36 and 60 and pl. 4.7), Dick Askew (pl. 2.4), Keith Bland (figs. 85–87), Sarah Bunker (figs. 31 and 32 and pl. 4.2), Brian N. K. Davis (fig. 37), Steven Falk (fig. 66 and pl. 4.1), Francis Gilbert (fig. 30), Miranda Gray (fig. 92 and pls. 2.5–2.6, 3.1–3.3, 3.5, 3.6, 4.4 and 4.5), Peter Hayward (figs. 42–46, 48, 49, 51, 52, 54 and 64 and pls. 1.2–1.6, 2.1, 2.2 and 2.7–2.9), Anthony J. Hopkins (fig. 90 and pls. 4.3 and 4.6), William Kirk (fig. 84), Margaret Redfern (fig. 38), John Rodford (fig. 59 and pl. 2.3), Keith Snow (fig. 65), Jo Wright (figs. 13, 17, 18, 20, 23, 33, 35, 40, 69–80, 83, 88, 89 and 91) and Dennis Unwin (figs. 4 and 5).

Most of the statistical tables were calculated using the algorithms in Microsoft Excel 2000. Tables A3, A4 and A6 are extracted from Neave, H. R. (1995) *Elementary statistics tables*. Routledge, London (pages 29, 30 and 40), and Neave, H. R. (1995) *Statistics tables*. Routledge, London (tables 5.1 and 5.3), by kind permission of the publishers.

C.P.W.
P.A.C.

1 Designing an investigation

The aim of the Naturalists' Handbooks is to guide those interested in studying natural history. To this end the books in this series introduce the natural history of a group, and provide identification keys, ideas for research and details of some relevant techniques. This volume is designed to complement others in the series by giving a general overview of techniques and methods of analysis used in ecological sampling and investigation, to support readers who wish to undertake research of their own.

Much can be learned from simply observing plants and animals, in the field or in a more controlled environment such as a greenhouse, and seeing how they grow, feed, reproduce, and interact with their surroundings. Describing these processes is often an important starting point in any understanding of natural systems. However, instead of making observations on a single individual, we often wish to examine a group of individuals. Now the information gathered becomes more complex, and in order to record and understand what is going on it may be necessary to summarise or analyse the data in some way. This book introduces ways of designing and analysing experiments and surveys so that complex situations can be described and summarised, comparisons can be made, and interactions between organisms and their environment can be examined objectively. We refer to a wide range of possible investigations, many of which have been suggested in other books in the series, which give relevant background information. Although we concentrate on sampling and surveying invertebrate animals, some techniques (especially the statistical methods described in Chapters 4 and 5, and the advice on displaying data in Chapter 6) will be applicable to a wider range of topics. In addition, we discuss how to monitor environmental variables and plants, as well as invertebrate animals (Chapter 2), and how to begin to identify invertebrate animals (Chapter 3).

To start with an example, if we wish to investigate the types of animals found inhabiting a nettle patch, we might record how many individual animals there are, or the numbers of animals of each feeding type (such as predators or herbivores), or how many there are in each taxonomic unit (such as beetles, spiders or flies). Such counts give us a picture of the animals found in one nettle patch. But our chosen nettle patch might be unusual in some way, so to get data that are representative of nettle patches in general, we could obtain counts for several nettle patches. Now we need some descriptive statistics to summarise the animal counts across all the patches. Techniques for summarising this type of information are examined in Chapter 4.

On the other hand our requirements might be more sophisticated. Broadly, there are three main types of questions that we might ask, exemplified here in relation to an investigation designed to see whether and how the numbers of animals found under decaying logs on a woodland floor are influenced by the size of the log.

1. Is there a difference between samples? For example, if we divide the logs into two classes, large and small, are more animals found under logs of one size class than another?

2. Is there a relationship between variables? For example, if we count the animals under logs of a wide range of sizes, do the numbers of animals vary in some systematic way, either increasing or decreasing, as log size increases?

3. Is there an association between frequency distributions? For example, if we divide the animals found into predators, herbivores and decomposers, is there an association between the size class of logs and the frequency of one or more feeding type of animal?

We will examine each type of question in more detail later in this chapter. The analytical techniques needed to answer these questions will be considered in Chapter 5.

Designing and setting up experiments and surveys

It is important to think about how your results will be analysed before beginning to collect data, because a given technique may require the data to be collected in a particular way. Therefore, when an experiment or survey is set up, the analysis must be built into the initial design.

The main types of investigation used in research are experiments and surveys. The distinction between these is important: an experiment involves manipulation of a system, whereas a survey involves making observations without manipulating the system. For example, to discover how many insects could be found under logs of various sizes, we could either survey a woodland floor finding as many logs as possible and recording both the number of insects and the size of the log, or we could devise an experiment in which we placed logs of different sizes on a woodland floor and after a period of time counted the number of insects underneath each one. The advantage of the experimental approach is that we can standardise all aspects of the logs except size. For example, we might standardise the degree of decay, the type of wood, and the distance from the nearest log. All of these factors might influence the invertebrates found and confuse any relationship with log size. A survey has different advantages in that it gives an impression of what is happening in real life, under logs that have been deposited naturally. Moreover, the survey approach is less likely to damage the environment. In most environmental research programmes, surveys are useful for generating ideas about possible influences, but because of the additional complexity in real situations, they cannot identify cause and effect. Experiments strip away the additional complexity, so they are more useful in identifying direct influences, but they are sometimes less relevant to real situations.

Types of data

In designing an experiment or a survey, it is necessary to think carefully about the type of data to be collected. The features that are recorded (such as heights of plants, ages of trees, numbers of insects, densities of animals per plant) are termed variables. A variable can be one of three types. The simplest are categorical (or nominal) data in which each value is recorded as one of several distinct categories (such as purple or yellow flowers; predators, herbivores or decomposers). Where the categories can be ranked in some kind of logical order (such as large, medium or small trees; urban, suburban or rural habitats) the data are described as rankable (or ordinal). The most detailed type is measurement (or interval/ratio) data. Not only are these rankable, but also there is a known interval between adjacent items in the sequence. Examples of measurement data are the numbers of soldier beetles on thistle flowerheads, temperatures in the centre of patches of plants or the heights of trees in metres.

Sampling strategies

It is rarely possible to collect information on every single animal (or log, nettle patch or plant) present. We often need to examine a smaller number of items which we hope are representative of the situation as a whole. The total number of data points that could theoretically be gathered is known as the population; the actual number used is a sample. Sampling is used in many types of survey. For example, pollsters obtain opinions from large samples of people and use these to indicate the wishes of the population as a whole. Sampling must be planned with care. The larger the sample, the better it will represent the population. There are mathematical methods for working out how large a sample should be (Elliott, 1977*), but unfortunately they require a knowledge of how variable the system is: something that we do not usually know at the start of an investigation. A small pilot study will often reveal how variable the system is. If it is suspected, or known, that the system is very variable, take a large sample (over 50 data points). Otherwise aim for as large a number as is practical, taking into account constraints such as time, work force, equipment, or availability of material in the system.

Calculated aspects of a sample, such as the mean (average) number of hoverflies visiting flowers per hour, are called statistics. They are estimates of the true attributes of

* References cited under authors' names in the text appear in full in References and Further Reading on p. 110.

the population which are called parameters. All populations have inherent variation. Think about how many different opinions a market researcher could collect. Similarly, even on two very similar nettle patches there will be differences in the numbers of invertebrates found. To take this variability into account, sufficiently large numbers of items must be sampled, and care should be taken to ensure that the samples are not biased. For example, samples of the insects on thistle flower heads will give a biased impression of species composition if the samples are collected in the early morning, because any animals that are active later will be missed. If two areas are being compared, sampling one site early and one site later in the day will introduce another variable (time) into the comparison. We would be looking not only at two sites, but also at two times of day, making it impossible to separate the effects of the two variables or draw any conclusions. We could take into account the time of day by taking a series of collections alternately from our two sites, sampling first one site then the other, then back to the first, so that there was a spread of samples for each site over the day. Or we could take samples on successive days, reversing the order in which we sampled the sites. In situations where confounding variables (such as time in this example) cannot easily be taken into account, experiments come into their own. It is possible to design an experiment in which only one variable is manipulated whilst the others remain constant.

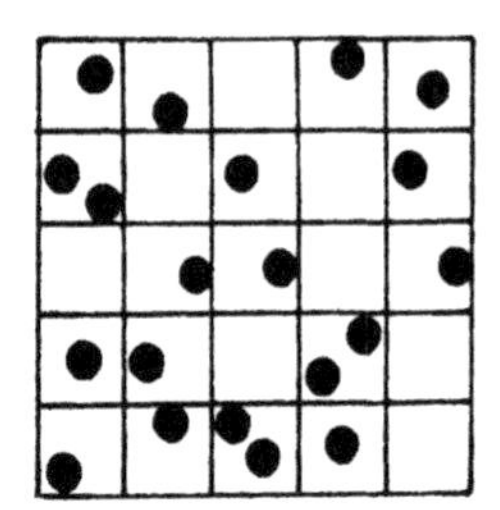

Fig. 1a. Randomly distributed sampling points.

There are several sampling layouts that help us to avoid bias. The first of these is random sampling (fig. 1a). Here random numbers are used to determine the order in which to sample plants, experimental plots, or survey sites. For example, to randomly sample 1 m x 1 m plots in a field, coordinates are used to position the sampling sites using pairs of random numbers generated by a calculator or computer, or taken from a table such as table A1 in Appendix II . If sites are first allocated numbers, random sampling may also be used to determine which site is to be visited first.

1b. Regularly (systematically) distributed sampling points.

An alternative strategy, systematic sampling, is an objective method of identifying the sampling points (fig. 1b). For example, to systematically sample the insects on nettle plants, we might collect from every tenth nettle in a clump. Usually this provides an unbiased sample of individuals unless for some reason every tenth nettle is more likely to show certain characteristics. Suppose we used systematic sampling to examine the distribution of ants' nests in grassland. We could place 2 m x 2 m quadrats (square sampling frames) 10 metres apart across the site and count the nests within each quadrat. However, if ants' nests are in competition with each other, they are likely to be spaced out (Skinner & Allen, 1996, NH 24*, give more information about ants' nests). If the nests happen to be about 10 m apart, we would either overestimate the number of nests if our sampling sequence included the nests, or underestimate it if

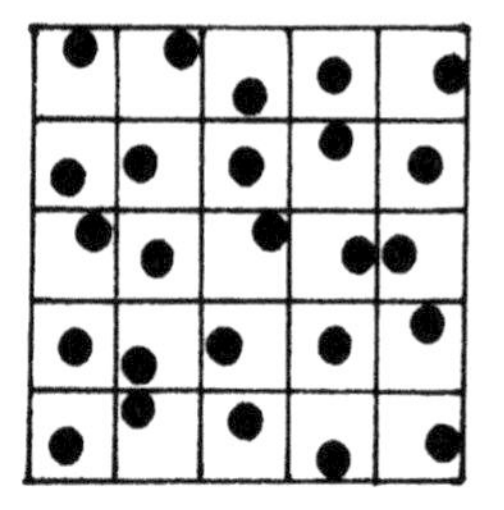

1c. Stratified random sampling points.

* References to Naturalists' Handbooks include a number indicating the number of that title in the series. For example, Skinner & Allen's *Ants* is no. 24 in the series.

the quadrats fell between the nests. In this situation it is better to use a mixture of random and systematic sampling (called stratified-random sampling) in which the area is divided into blocks (say of 10 m x 10 m) and the quadrats are placed randomly within each of these (fig. 1c). Using this technique, sampling can be distributed through time, as well as space.

Selecting independent data points

It is important to recognise sample designs that result in data being linked in some way. For example, we might compare the numbers of pond skaters found on the surface of ponds in the shade, with those on ponds exposed to the sun (Guthrie, 1989, NH 12, describes animals of the surface film). It is not appropriate to take measurements from just one pond in the shade and one in the sun. Although such a survey could produce many measurements, the data would be representative of only the two ponds sampled, and not of ponds in general. The several data points from each pond are linked by virtue of the pond from which they were obtained, and are likely to share many different attributes with one another. In this sense the data are not independent. Therefore it is preferable to take single counts from a large number of ponds in the shade and from a similar number of ponds in the sun.

It is usually better to collect approximately similar numbers of measurements from each sample. If we found over twenty ponds in the shade, but only ten sunlit ponds, we might be tempted to take twice as many measurements from each sunlit pond. However, here again, taking more than one measurement per pond involves non-independent data. It is better to take single measurements from approximately similar numbers of individual ponds in the shade and in the sun. If equal sample sizes are not easy to find, it is better to use unequal sample sizes (20 shaded and 10 sunlit ponds) than to use non-independent data. Few statistical tests require equal sample sizes. On the rare occasions where unequal sample sizes are unacceptable, it is better to reduce the number of shaded ponds sampled than to use non-independent data.

In some sample designs the data are linked to each other deliberately. For example, we might be interested in comparing counts of animals found on cabbages before and after the application of a pesticide, matching the two samples from each individual cabbage, or the numbers of mayfly larvae found above and below sewage outflows into a series of streams, matching the counts from each individual stream. These designs are perfectly sound, and allow us to use powerful matched-sample tests (Chapter 5).

Asking questions

There are various questions that we might ask as part of an investigation and it is important to be clear about possible methods of analysis in advance of any sampling.

Is there a difference between samples?

We often need to know whether two (or more) samples differ in terms of their component animals or other factors. For example, are there differences in the numbers of invertebrates in shaded and unshaded nettle patches? Here the numbers of invertebrates would be counted in a series of patches in the shade and in a similar number of unshaded patches. The sampling technique (and the effort put in) must be as similar as possible in the two types of nettle patch, as must all other characteristics such as patch size, plant height, location, and time of collection. The techniques for analysing such questions are given in Key A in Chapter 5 (p. 69).

The same sort of question might be asked about the number of invertebrates that are decomposers under logs that have fallen from oak, beech and pine trees (here differences between three samples are examined), or about differences in plant height between individuals of a particular species growing at the edges of woodlands and those in the centre.

For each data point we have two pieces of information: categorical data identifying which sample it is from (for example sun or shade), and a measurement (such as the number of invertebrates). It is also possible to examine differences between ranked data, such as the amount of light (the ranked variable) getting through the canopies of deciduous and coniferous woodlands (the categorical variable), where the light levels are scored on a ranked scale (for example from 1 to 10, where 1 is very dark, and 10 is very bright).

Is there a relationship between variables?

A second common research question is whether two (or more) variables are related. We might wish to know whether the numbers of insects found on a series of dock plants are related to the heights of the plants. For each plant we would measure the height and count the number of insects. Again, other factors (whether the plants are in sun or shade, and the time of day at which the collections are taken) should be kept as consistent as possible. Now we have pairs of measurement data (the height of the plant and an associated count of animals for each plant) that can be examined using the techniques described in Chapter 5 (Key B) (p. 84). Similar questions could be asked if one or both of the variables were ranked, for example in the relationship between the colour of cabbages (ranked on a scale from light to dark green) and the numbers of aphids found on the plants. An extension to relationship testing involves producing simple models from the data in order to be able to predict one variable from another. The variable that is likely to be influenced by another is termed the dependent variable, whereas the variable that is likely to be the influential one is termed the independent variable.

Is there an association between frequency distributions?

A third common type of question relates to frequency analysis. For example, an examination of the frequency of green and brown individuals of a particular insect species found on green and brown plants might reveal an association between animal and plant coloration, possibly related to camouflage and the incidence of predation. Note that for each observation there are two categorical variables (whether the insects are green or brown, and whether the plants are green or brown). In this case, the frequencies in each category (the numbers of green insects on green plants, the numbers of green insects on brown plants, the numbers of brown insects on brown plants and the numbers of brown insects on green plants) would be compared using the techniques discussed in Chapter 5 (Key C) (p. 89). Such questions can easily be extended, for example by increasing the number of categories of insect colour, plant colour or both.

Another type of frequency analysis concerns the difference between observed frequencies and an expected outcome. For example, we might want to know whether we are more likely to find male or female spiders sitting in webs within thistle plants. Without any prior knowledge, it might be reasonable to assume that there should be approximately equal numbers of males and females, and any deviation from a 50:50 split would be of interest. In both types of frequency analysis, we are looking at the frequencies of categorical variables.

2 Sampling invertebrates

Before beginning a sampling programme it is important to be sure of one's aims, and to consider the principles outlined in Chapter 1. Sampling methods include recording whether or not certain species are present, estimating population size, and examining community structure by establishing the relative abundance of each species. A wide range of techniques can be used to sample invertebrates. Understanding the basis of each method allows us to assess whether it is appropriate for a particular investigation. The following sections deal with some common methods grouped according to the habitat in which the invertebrates are most active (for example, on plants, in the soil, in the air). To take account of all life stages of the animals (for example, eggs, larvae, pupae, and adults) it may be necessary to examine several different habitat types and to use a different technique for each life stage. For example, dragonflies, mayflies and mosquitoes have aquatic larvae but flighted adults, so we may need to survey the depths of ponds for larvae and the water surface and emergent vegetation for egg-laying and mating adults, while more terrestrial habitats may be used by adults searching for food. Miller (1995, NH 7), Harker (1989, NH 13), and Snow (1990, NH 14) give more information on dragonflies, mayflies and mosquitoes respectively.

Direct monitoring of the abiotic (non-living) elements of the environment may be a complex task. There is increasing interest in the use of biotic (living) indicators of some environmental factors. Many such indicators are species whose distribution depends on their tolerance of certain environmental factors, such as pollution levels. Commonly used indicator groups include freshwater invertebrates for organic pollution (Hellawell, 1986; Mason, 1996) and lichens for sulphur dioxide levels (Richardson, 1992, NH 19). Several groups of insects have shown promise as indicators of habitat quality including ground beetles and hoverflies (Speight, 1986). Both groups are ubiquitous, relatively easily identified, and well documented so we have a wealth of published information (for ground beetles see Forsythe, 2000, NH 8, and for hoverflies see Gilbert, 1993, NH 5).

In studies of invertebrate animals, it is often useful to examine the structure and diversity of a habitat. Texts that provide background ecology of specific habitat types include Wheater (1999) for urban habitats, Read & Frater (1999) for woodlands, Fielding & Howarth (1999) for uplands, Price (2003) for grasslands and heaths, and Dobson & Frid (1998) for aquatic habitats. In addition to documenting the invertebrates themselves, it may be necessary to record aspects of the vegetation, the microclimate, and properties such as pH, stream flow and soil texture. This chapter deals first with methods for surveying vegetation and the physical and chemical aspects of the environment, and then gives examples of techniques involved in sampling invertebrates.

Conducting fieldwork responsibly and safely

When planning fieldwork it is important to take into account your responsibilities and any legal implications of the work. Look after your own health and safety and that of those around you, and try to avoid adversely influencing the environment. Always seek permission to work on a site before you start the fieldwork. Keep disturbance to a minimum, and remove as few samples as possible. Identify specimens *in situ* if you can so that they need not be removed from the habitat. Whole plants may not be taken without the express permission of the landowner. Some species are protected in law and may not be removed or even disturbed. Examples include most birds, badgers (and their setts), some small mammals (such as shrews), great crested newts, bats, and some wild flowers (including many species of orchids). In addition, rare species should not be taken unless absolutely necessary. Guidance may be obtained from a national agency, such as English Nature (address p. 97).

The project should be assessed for any risks, including those caused by the terrain, the techniques and any sudden changes in weather. Any chemicals being used should be checked against COSHH regulations (Control of Substances Hazardous to Health Regulations, 1988) in terms of use, safe disposal and how to deal with spillages (The Health and Safety Executive provides advice, address p. 98). Many organisations have their own health and safety guidelines. In the absence of these, advice is available in publications such as those of Nichols (1999) and Piggot & Watts (1996).

Try to avoid working alone in the field. If you must work alone, check out and back in with someone who knows your planned routine. Clothing and footwear should be suitable for the terrain and climatic conditions (warm and waterproof). Safety glasses and gloves should be worn to handle chemicals, and suitable gloves can offer protection against thorns and infection from soil- and water-borne disease. Keep tetanus injections up to date, and take particular care where there is risk of Weil's disease (near ponds, rivers and canals) and Lyme disease (which is transmitted by ticks). When working away from the base, carry a first aid kit, emergency food, maps, a torch, mobile phone, whistle, compass, watch and survival bag. In general, avoiding risks will help to ensure problem-free fieldwork.

Surveying vegetation

The defining characteristic of most terrestrial habitats is its vegetation. This often provides the structural components in and on which invertebrates live. It is, therefore, often necessary to identify plant species and communities, and also to measure characteristics such as cover, plant height, and community structure.

Plant cover is usually assessed by quadrat sampling. Quadrats (square frames, fig. 2) are laid out, each species within is identified and its percentage cover estimated. The size of quadrat used depends on the nature of the habitat: 0.5 m x 0.5 m or 1 m x 1 m frames are often used in grasslands, whereas larger quadrats (2 m x 2 m) are better for woodlands. Smaller quadrats (0.1 m x 0.1 m) may be useful for smaller plants (such as lichens and mosses). Estimation of cover is easier if each quadrat is subdivided into a grid. For example, a 0.5 m x 0.5 m quadrat can be divided into 25 squares each 0.1 m x 0.1 m by taut strings attached to the frame. It is often easier to estimate percentage cover by scoring the presence or absence of each species in each grid square than to count the numbers of individual plants, since it may be difficult to separate individual plants by inspection of the above-ground structures. For example, in clumps of bracken, the fronds are often connected below ground so that a single plant forms a large patch. Grasses are also difficult to separate from each other. Standardised scoring systems may be used instead of estimates of percentage cover. Two common systems are the Domin and the Braun-Blanquet scales (table 1). These provide fixed reference points for a range of percentage covers. A simpler method that provides a rankable scale is the DAFOR system in which each species present is allocated a code based on its abundance: Dominant, Abundant, Frequent, Occasional, Rare.

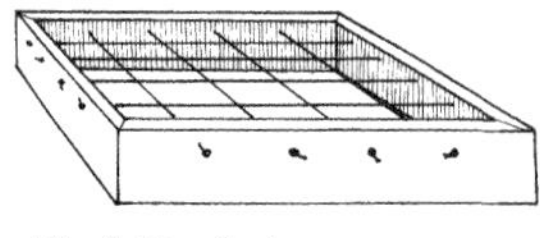

Fig. 2. Quadrat.

Table 1. *Braun-Blanquet and Domin scales.*

Braun-Blanquet scale	**Value**	**Domin scale**	**Value**
Less than 1% cover	+	Single individual with no measurable cover	+
1–5% cover	1	1–2 individuals with normal vigour and no measurable cover	1
6–25% cover	2	Several individuals with under 1% cover	2
26–50% cover	3	1–4% cover	3
51–75% cover	4	5–10% cover	4
76–100% cover	5	11–25% cover	5
		26–33% cover	6
		34–50% cover	7
		51–75% cover	8
		76–90% cover	9
		91–100% cover	10

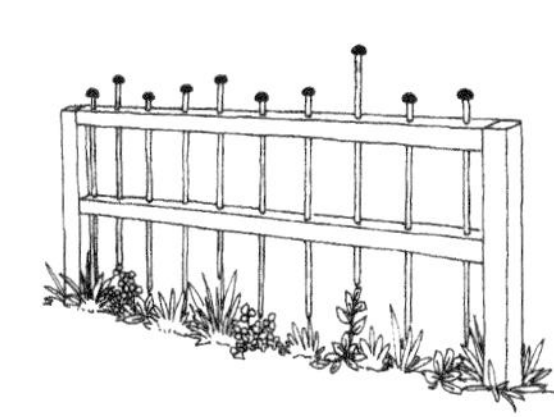

Fig. 3. Pin frame.

Quadrats can be positioned randomly, using random number tables (table A1, Appendix II) to select their coordinates, or more systematically to ensure that all areas in a site are covered or to measure vegetation at varying distances from some feature such as a footpath or field edge. Alternatively, they may be positioned in a stratified random way (see Chapter 1). Often quadrats are positioned along lines called transects. Belt transects are continuous lines of quadrats. Ladder transects have gaps between adjacent quadrats allowing larger areas to be covered with no increase in sampling effort. The line transect method does not employ quadrats. Instead plants touching the line at certain positions (for example, every 1 m) are recorded. This approach can be used for rapid sampling of the frequency of species found in a site, but it gives little information about vegetation cover. An alternative to quadrats that allows estimations of percentage cover is the pin frame (fig. 3). A series of long spikes (similar to knitting needles) are arranged at, say, 0.1 m intervals along a vertical frame and the species of plant touched by each point is recorded. There are usually ten pins in a frame. The proportion of pins touching a particular species indicates the percentage cover of that species.

Additional measurements that can help to characterise the vegetation might include plant height and width, tree girth, density, age structure, condition, architecture, and disease incidence. Age of trees may be assessed from cores taken from the trunk using purpose-made tree corers, while factors such as condition and disease incidence may be scored on a ranked scale of 1 to 5. Tree architecture may give clues about the historical ecology of a woodland. Straight trunks with no sign of old branch scars indicate that the trees grew up in relatively dense woodland and did not spread their branches until they reached the canopy. Trees with low spreading branches probably grew in the open or in large gaps, and those with signs of old branch scars low down may have grown in a gap which then closed, resulting in the loss of lower branches and concentration of foliage in the canopy layer.

There may be more than one layer of plant cover. This is easiest to see in a woodland where there may be a ground flora, with shrubs and tall herbaceous plants above. Yet higher there may be a subcanopy layer of woody plants, overlaid with a canopy through which a few tall emergent trees grow. Even in grasslands, there may be layers of mosses, overtopped by grasses, with taller herbaceous and some woody plants emerging. Estimating the cover separately in each layer may reveal factors that influence the range and type of invertebrates present.

Books that cover the range of plant survey methods include a summary by Bullock (1996) and more detailed discussion by Moore & Chapman (1986) and Kent & Coker (1992). Books for plant identification include: Purvis and others (1998), Dobson (2000) and Richardson (1992, NH 19) for lichens; Courtecuisse & Duhem (1995) for the larger fungi; Watson (1981) for mosses; Merryweather & Hill (1992) for ferns; Phillips (1980) for common species of grasses, ferns, mosses and lichens; Fitter and others (1984) for grasses, sedges, rushes and ferns; and Hubbard (1984) for grasses; Fitter and others (1996), and Rose (1981) for flowers (the latter also includes a key to plants not in flower); and Mitchell (1978), and Mitchell & Wilkinson (1991) for trees. Stace (1997) is a specialist text covering all grasses, ferns, trees and flowering plants native or naturalised in Britain.

Measuring environmental properties

Animals are influenced not only by vegetation, but also by the physical and chemical characteristics of the environment. These include microclimate (localised temperature, incident radiation, humidity, wind speed), soil characteristics (chemical and physical properties including pH, texture, water content), and, for aquatic animals, the chemical and physical attributes of water (depth, flow, oxygen content), as well as levels of pollution such as sewage and of toxins such as heavy metals and pesticides. Monitoring some environmental properties such as pollution levels usually requires sophisticated equipment, hence the interest in potential biological indicators (p. 8).

Measuring microclimate

The microclimatic conditions affect the development and survival of invertebrate animals, and even small differences in vegetation height or density may have large effects on features such as temperature and humidity. Small invertebrate animals generally have a body temperature close to that of the immediate microenvironment, and the ambient temperature dictates when they can move around to forage and find mates. Larger invertebrates lose heat relatively slowly and may either bask or produce heat internally during exercise to raise their body temperatures above ambient. Although apparatus for monitoring microclimate is often expensive, useful equipment can often be bought or built quite cheaply. Here we outline some of the cheaper methods, further details of which can be found in Unwin (1978, 1980) and Unwin & Corbet (1991, NH 15) who explain how to make and use them.

Even a crude measure of temperature can help us to interpret fluctuations in the activity of animals. Normal mercury thermometers can be used to measure water and soil temperatures. For the latter they are set into brass tubes and pushed gently into the ground. Mercury thermometers are less suitable for measuring air temperatures because they are so big, and take so long to equilibrate, and above all because they over-estimate air temperature if used in the sun. If they must be used for measuring air temperature, a simple shield made from aluminium foil will protect them from sun, wind and rain and help to prevent spurious readings (fig. 4). Maximum/minimum thermometers enable daily fluctuations to be kept under review. Smaller sensors

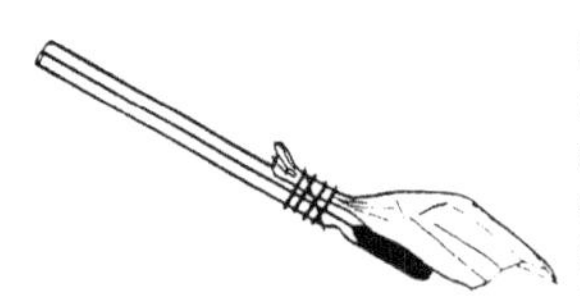

Fig. 4. Shield for thermometer.

such as thermistors or thermocouples are required to measure temperatures in microhabitats such as under leaves, in flowers or under debris such as logs and stones. Fine thermocouples are easy to make, but they need to be plugged into a rather costly meter. Thermistors and the associated meters are relatively inexpensive and are available from suppliers such as RS Components (address p. 100). Data loggers that print out or record temperatures at regular intervals are available from several suppliers (such as Philip Harris, address p. 99), although they may be quite expensive.

Incident radiation can be very important to small animals, as well as to plants. It is easy to record whether the sun is out or not, and cloud cover can be estimated in terms of tenths of the sky covered by cloud. Light levels may be simply measured using a photographic light meter (Unwin & Corbet, 1991, NH 15), although this will not directly represent the wavelengths most important for plant development. Photosynthetically active radiation (PAR) can be measured using a quantum flux meter, or with a cheaper home-made solarimeter as described by Unwin & Corbet (1991, NH 15). When the sun is shining, black bodies (including bees and mercury-in-glass thermometer bulbs) absorb incident radiation and may become warmer than the air around them. This effect is important to insects, and it can be revealed by measuring black globe temperature as well as air temperature. A miniature black globe can be made conveniently by fitting a thermometer into a 9-mm diameter sphere of Blu-tack painted matt black (Corbet and others, 1993). The temperature of an unshaded black globe in sunlight is higher than that of a shaded air temperature sensor. Paired records of black globe temperature and air temperature can be used to give an indication of the timing of periods of sunshine.

Relative humidity is also important for many invertebrate animals. Those without a waterproof cuticle tend to lose water faster than those with more protection. The tendency for rapid water loss may be associated with characteristic behaviour, for example vulnerable animals may group together under stones and other debris during the day when the risk of drying out is highest. Positions on plants even a few millimetres apart may be quite different in the degree to which they expose animals to drying out. Unless basking, many insects avoid the upper surface of leaves during the day, especially in direct sunlight, whilst the underside of the same leaf, or adjacent leaves in the shade, may be covered with feeding animals such as greenflies. Relative humidity may be measured using simple hygrometers, but these respond slowly and are not suitable for small-scale measurements. Fine-scale instruments are commercially available but expensive. Unwin & Corbet (1991, NH 15) discuss the measurement of humidity and suggest ways of improvising inexpensive measuring devices.

Wind speed interacts with other aspects of microclimate, for example by disturbing moist air and providing a cooling effect in sunlit areas, and it also has a

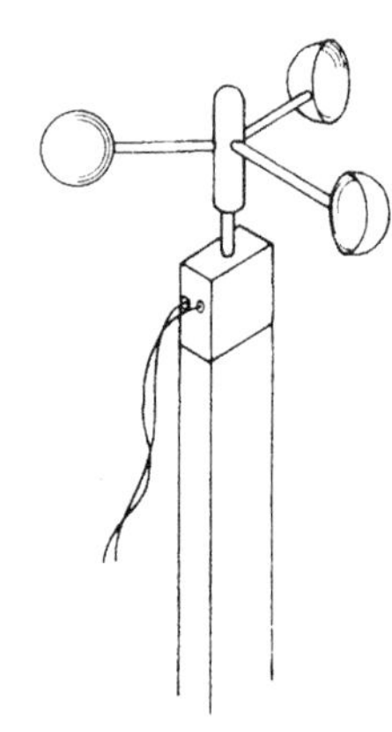

Fig. 5. Cup anemometer.

direct effect on many flying insects. Even a slight breeze may ground small species such as mosquitoes, while higher wind speeds prevent larger animals such as butterflies from flying. Some non-flighted invertebrates such as small spiders (under 1.6 mg) also make use of the wind to carry them long distances. A small spider crawls up a fence post or similar upright object, extrudes a length of silk from the spinnerets and is picked up by the wind. By releasing the silk line it can drop out of the airstream. Such animals will not attempt to disperse in this manner when the wind conditions are not appropriate (that is, when it is too calm or too windy). Wind speed is measured using an anemometer. The most familiar type, a cup anemometer, has three cups set on a shaft that rotates freely at a rate that depends on the wind speed (fig. 5). If the space in which wind speed is to be measured is too small to accommodate a cup anemometer, a hot-wire anemometer, which depends on the cooling action of wind on a heated wire, can be used (Unwin & Corbet, 1991, NH 15).

Monitoring substrates such as sand and soil

The nature of the substrate (such as sand or soil) has a major influence on the distribution and activity of invertebrates. The chemical and physical properties of soils affect invertebrates directly and indirectly, for example via influences on the vegetation. A major influence is the water content of the soil. Most of the available water is held by capillary forces within the soil structure. The more small channels there are, the better is the water holding capacity of the soil. Sandy soils consist of large particles and so have rather few, large channels which have weak capillary forces and allow much water to drain away leaving the soil quite dry. Sandy soils warm up quickly in the sun and cool down rapidly at night and, although they tend to be generally warm, they may have quite variable temperatures. Clay soils consist of very small particles and so have many tiny channels that hold water well. These soils are much wetter and heavy clays may become waterlogged. However, when they dry out they may become hard and sometimes form surface cracks. Clay soils are cooler and experience less extreme fluctuations of temperature. Other soil types such as loams show intermediate characteristics. Soil types may be identified using keys such as those in Brodie (1985) or Trudgill (1989). The water content is measured by weighing a sample and then heating it at about 105°C until a constant mass is obtained. At this point the water will have been driven off and the difference in the two mass measurements (usually expressed as a percentage of the initial mass) is the moisture content of the soil. Soil samples should be taken randomly or using some systematic approach (p. 4). Cylindrical plugs of soil are usually removed using specially designed soil augers (fig. 6), but bulb planters (from garden centres) provide reasonably standardised samples.

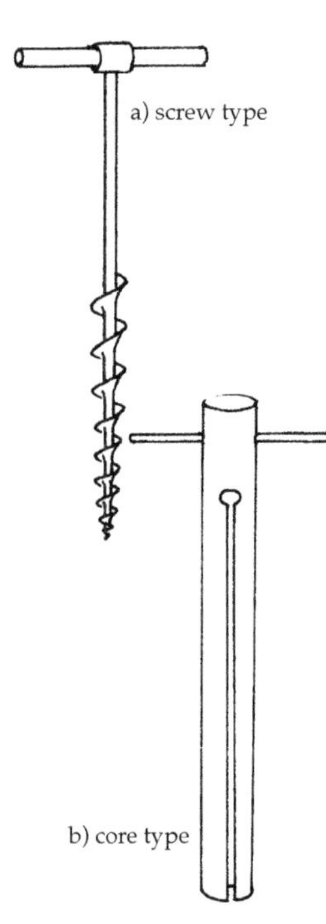

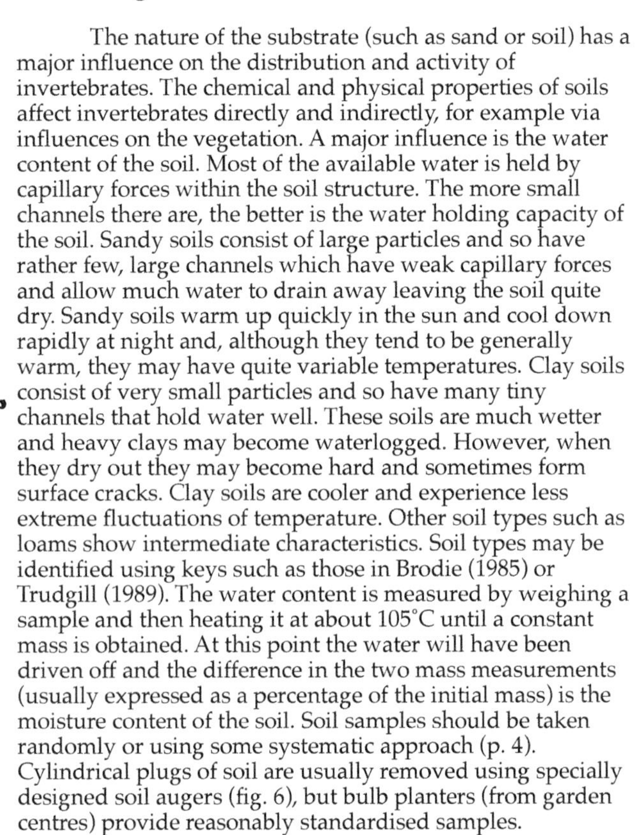

Fig. 6. Soil augers.

The nutrient status of the substrate is often an important consideration. Sophisticated equipment is available for measuring concentrations of major nutrients such as nitrogen, phosphate and potassium, but simple soil-testing kits (from garden centres), although less accurate, are much cheaper and easier to use, especially in the field. Such kits may also include the facility to measure soil pH. This assessment of the acidity or alkalinity of the soil can give clues about the chemistry of the soil. For example, in alkaline soils some nutrients such as phosphate are locked up as insoluble compounds, while in acid soils, other chemicals such as aluminium may be released in large, sometimes toxic, quantities. The amount of organic material in the substrate may influence not only the nutrient status but also the water holding capacity. Organic content can be measured by comparing the mass before and after carbon is burnt off at high temperatures in a special furnace. Approximations may be made by floating the organic material off in a bowl of water. The two fractions (that which sinks and that which floats) can then be dried to constant mass. The mass of the dried floating fraction (organic material) may then be expressed as a percentage of the total dried mass (of both fractions combined). Pollutants are generally assessed using expensive analytical equipment, but some soil-testing kits include facilities to test for contaminants such as lead and zinc.

Monitoring water

In aquatic habitats, chemical and physical factors that you may wish to measure include depth (using a metre rule in shallow water or a plumb line in deep water) and the flow rate of streams and rivers. There are purpose-built flow meters that can be used at several depths within the river, but one of the easiest ways to measure surface flow in long straight sections of a river is to time a suitable floating object as it moves between two points 100 m apart. A suitable object is one with a specific gravity similar to that of water so that it does not sink, but does not float too high out of the water. An orange is ideal.

The oxygen content of the water determines the range of organisms that can survive. It is possible to buy a dissolved oxygen meter (for example, Philip Harris, address p. 99) or to use a chemical titration technique such as the Winkler method (Stednick, 1991). Where the water is polluted with organic matter such as sewage or farm waste, the processes of decomposition may remove oxygen faster than it can be replaced. The resulting biochemical oxygen demand (BOD) is measured by incubating the water for five days at 20°C and measuring the oxygen used during the period. Other important pollutants include heat. Thermal pollution occurs, for example, when water is extracted from a river to be used as a coolant in a factory and is then reintroduced later several degrees warmer. A rise of water temperature reduces the oxygen-holding capacity of water, although the turbulence often associated with extraction and reintroduction may increase the interchange of oxygen between the air and the water. Water temperature may be measured using simple thermometers or thermocouples.

The nutrient status of waters may indicate exposure to organic pollution. Eutrophication (that is, nutrient enrichment, often by nitrates and phosphates) can stimulate algal blooms which, on death and decomposition, may cause serious depletion of oxygen in the water, leading to the death of fish and invertebrates. As well as measuring the BOD, one can examine nitrate and phosphate levels using colour comparator kits (Philip Harris, address p. 99) or by spectrophotometry. Nitrate is only one of several sources of nitrogen that may be found in waters. Others are ammonium (often due to agricultural runoff from fertilisers, or from domestic pollution) and nitrite (an intermediate product in the natural conversion of ammonium to nitrate). Separate analyses are required for each type to assess the total nitrogen level. Phosphate also exists in several forms in waters. These can be present in either dissolved or particulate forms, which may be examined separately by filtering the water through a 45 μm filter. Dissolved phosphate can be measured from the filtrate, and the amount of particulate phosphate obtained by subtraction of dissolved phosphate from a measurement of the total concentration. Stednick (1991) gives further details on the chemical analysis of water samples. Other contaminants such as heavy metals or agrochemical runoff may require more sophisticated analysis (Allen, 1989).

One form of pollution that can have a serious impact on invertebrate communities, and which is readily monitored, is turbidity. Suspended solids increase the cloudiness of the water, inhibiting photosynthesis below the surface, and so reducing oxygenation of the water. In addition, suspended solids may clog the feeding apparatus of filter-feeding invertebrates such as freshwater mussels or water fleas. Spectrophotometry can be used to obtain absolute measurements of the amount of suspended material in water samples, but a simple comparative method of assessing turbidity uses a Secchi disk. This is a disk, 20–30 cm in diameter, coloured white (for marine work) or with alternate black and white quadrants (fig. 7). The disk is lowered slowly into the water and the depth at which it disappears from view is recorded. It is then raised slowly and the depth at which it reappears is also recorded. The mean of these two depths, called the Secchi disk depth, is a useful measure of the turbidity of the water since the disk will be visible at greater depth in clear water than in cloudy water.

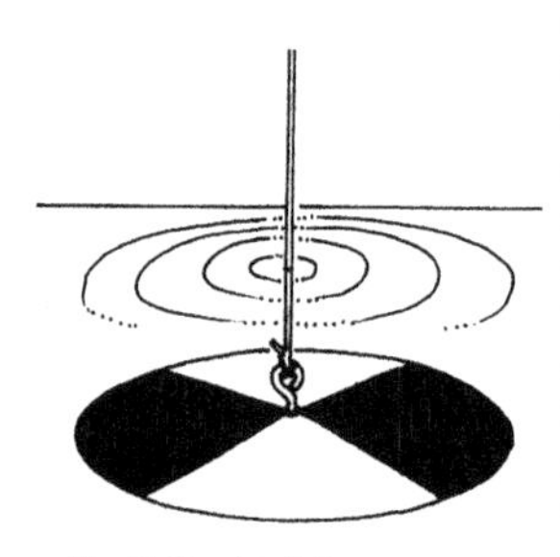

Fig. 7. Secchi disk.

Salinity is important in determining which organisms can survive in brackish water, for example in estuaries. It also becomes important in freshwaters that are subject to saline pollution, for example from runoff from roads following de-icing in winter. The distribution of species along a salinity gradient within an estuary or other brackish water habitat can be an interesting avenue for investigation. McLusky (1990) gives more information about estuarine habitats. Salinity can be measured with a salinity meter, or by a chemical method of chloride determination (Stednick, 1991, gives further details). Alternatively, salinity is sometimes estimated by measuring the conductivity of the water using a conductivity meter. This assesses the electrical resistance of the water sample, which is related to the total concentration of dissolved substances (particularly inorganic compounds) in the water. If salt is the major dissolved substance, conductivity can provide a reasonable index of salinity differences between samples, but if other ions (for example from agricultural runoff or sewage) are also present in large quantities, they will contribute to the conductivity which cannot therefore be taken to reflect salinity alone.

Ideally, physical and chemical properties of water should be measured in the field as soon as the sample has been taken. If field measurements are impossible, there are some preservation methods which allow water to be held in storage. The maximum period following preservation for which samples can be kept before analysis depends on the factor being determined, but varies between a few hours for BOD to six months for metal concentrations. Stednick (1991) gives further details of appropriate preservation methods. If it is necessary to bring water samples back to a laboratory for analysis or preservation, they should be packed in cool boxes (preferably at about 4°C) and kept in the dark. Sample bottles must be filled with water completely and sealed, so that there are no air bubbles.

Measuring physical attributes of a habitat

Other physical components include habitat size, aspect, topography and altitude. The size of the habitat block may influence the ease with which animals colonise and the numbers of animals it can support; larger habitats are easier to find and tend to support more animals. Habitat size can be measured directly (for example, the width of a stream or river, the length of a hedgerow, or the size of a log) or indirectly from maps (for example, the area of a lake or a woodland). In the northern hemisphere, south-facing habitats are exposed to more sunlight and tend to be warmer and drier than north-facing habitats. Aspect often influences the type of vegetation and the animal life. For example, the southern slope of some ants' nests is shallower than the other slopes, gaining maximum advantage from the sun's rays at noon (Skinner & Allen, 1996, NH 24). Many insect eggs have shorter development times at higher temperatures, for example those of grasshoppers (Brown, 1990, NH 2) and solitary wasps (Yeo & Corbet, 1995, NH 3). Conversely, the queens of some bumblebees seek out north-facing sites in which to overwinter (Prŷs-Jones & Corbet, 1991, NH 6). Interesting comparisons could be made between the communities on warmer southern slopes and those on cooler northern slopes. Ways of measuring incident radiation are described on p. 13, while aspect is easy to measure (with a compass).

As well as affecting microclimate, topography may influence animals and plants in other ways. Undulations in the ground can create differential drainage patterns which result in heterogeneity in plant cover. Steep slopes may be unstable, especially when they lack vegetation. Slope angles can be measured by the simple surveying technique of sighting horizontally from a known position on one of a pair of calibrated posts to another a set distance away and reading the difference in height (fig. 8).

In a research report it is important to record the location of the study site. Maps can provide this information in the form of grid references. Global positioning systems (GPSs) are becoming relatively cheap and accurate and use satellite information to identify both grid reference and altitude. The accuracy of a GPS depends on the number of satellites from which it can take readings. Many GPSs have a facility to record this accuracy in terms of the reading plus or minus a number of metres. Altitude may influence exposure and microclimate and will therefore affect animals and plants.

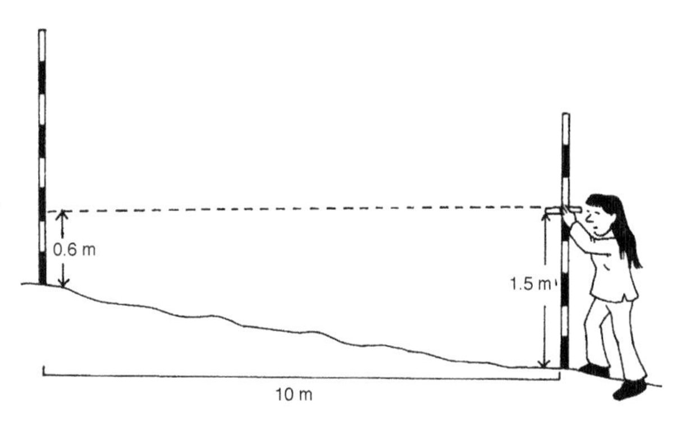

Fig. 8. Slope angles can be estimated by taking the inverse tangent ($\tan^{-1}$ on a calculator) of the ratio between the vertical distance (1.5 – 0.6 m) and the horizontal distance (10 m). That is, the angle here $= \tan^{-1}\left(\frac{(1.5 - 0.6)}{10}\right)$

$= \tan^{-1}(0.09)$

$= 5.1^{\circ}$

Sampling invertebrates from water

Unlike many terrestrial habitats, aquatic habitats often contain active invertebrates throughout the year. There are several layers within an aquatic environment: the surface film (Guthrie, 1989, NH 12); the plant communities growing in and beside the water (for example, Hingley, 1993, NH 20); the water body itself; and the substrate at the bottom of the water column. Each microhabitat requires a different sampling technique.

The water surface is often rich in resources as a result of the freely available oxygen and the tendency of materials to collect at the surface of water. Accordingly, the surface film supports a wide range of animal and plant groups (Guthrie, 1989, NH 12). The microscopic fauna and flora can be

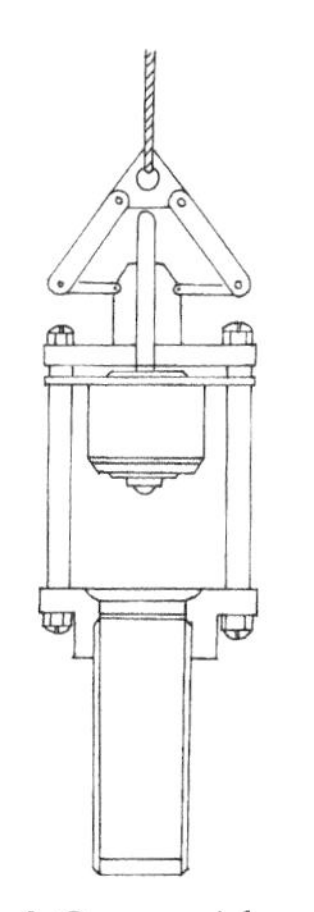

Fig. 9. Commercial corer.

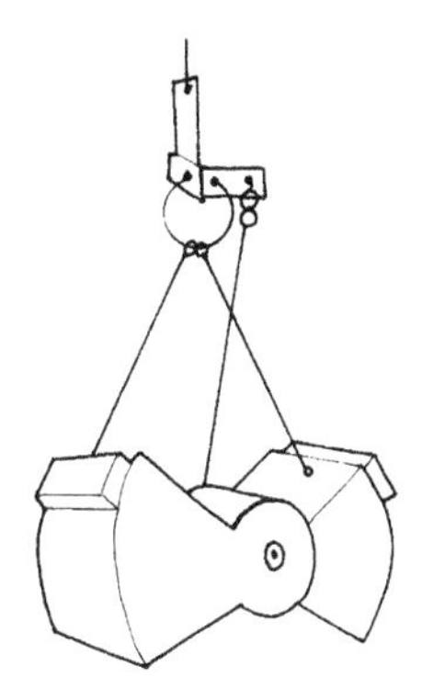

Fig. 10. Grab sampler.

Fig. 11. Kick sampling.

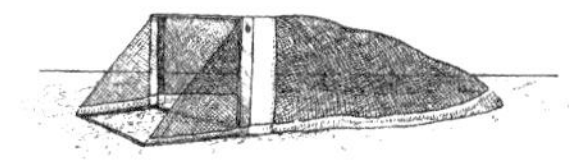

Fig. 12. Surber sampler.

collected by inserting a sheet of glass about 30 cm square through the surface film and withdrawing it slowly. The film that adheres to the glass can be removed using a windscreen scraper. Alternatively part of the surface film can be sucked into a pipette or syringe. Small floats of polystyrene, cork or cellulose left on the surface film will be colonised in a few days by the microfauna and can be squeezed out into collecting jars. All such communities require microscopic examination (Guthrie, 1989, NH 12). Insects flying just above the surface can be collected using long-handled butterfly nets. Larger animals living on or just under the surface may be sampled using pond nets swept through the surface layers, or by sucking them up in a pooter (p. 24) or ladling them up in a white plastic spoon. Alternatively, it is possible to mimic the surface film in some respects using water traps (p. 25).

Species that swim freely in the water column can be collected by sweeping with a pond net, or by dragging a plankton net slowly and steadily through the water. Animals living in and on aquatic vegetation are collected with a pond net or by hand searching, or by removing samples of the vegetation for sorting later.

Species living in and on soft substrates at the bottom of the water column can be sampled by scooping up silt and mud in one-litre plastic jars if the water is shallow enough, or by using corers in deeper waters. Silt and mud corers are plastic tubes about 0.5 m long and 5 cm in diameter with a valve at the top (fig. 9). They are lowered with the valve open and sink into the substrate. When they are pulled out, the valve closes trapping a core of sediment. An alternative can be improvised using a length of plastic drainpipe closed at the top with a tennis ball. In coarser substrates such as sand and gravel, a grab sampler is used (fig. 10). This comprises a pair of heavy jaws which sink under their own weight. When pulled back up the jaws close and a sample of sediment is captured between them. In all cases, the animals are separated from the substrate by gentle sieving, sometimes after dislodging them by a jet from a hose, or by pouring a suspension of sediment into a flat white tray and searching by hand. In shallow running water on stony substrates, kick sampling is effective (fig. 11). Stand in the water facing downstream with a flat-bottomed pond net held directly in front of you, with the open face towards you. By slowly but vigorously shuffling your feet you disturb the substrate and animals clinging to the stones are washed downstream into the net. A similar technique involves a Surber sampler which includes a quadrat within which the stones are disturbed, thus quantifying the area from which animals are collected (fig. 12). Mason (1996) gives more details about sampling animals living on the bottom of the water column. Colonisation samplers are useful in sampling animals living on the surface of the bed. These are left *in situ* for up to four weeks during which time the build up of organic material within them (including bacteria, protozoa, algae and organic debris) provides food for grazers such as snails or insect larvae. Standard samplers (called Standard Aufwuchs Units) can be made from double bundles of perforated black plastic cylinders with internal struts (seven in each tier) originally used in sewage treatment (Lee & Corbet, 1989).

Sampling invertebrates from substrates such as soil, leaf litter, sand and mud

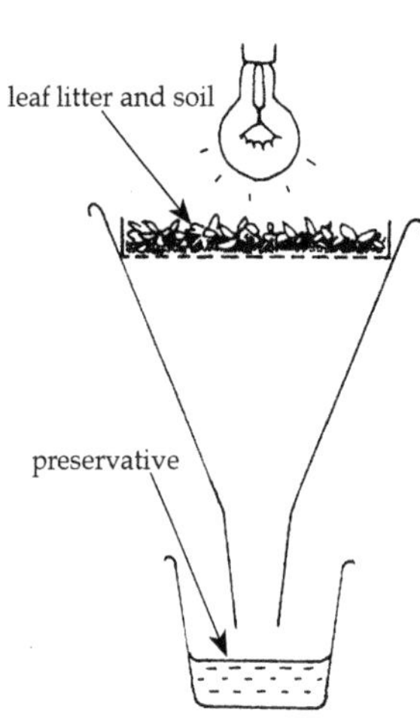

Fig. 13. Tüllgren funnel.

It is sometimes difficult to extract invertebrates from soil and similar substrates because many live at depth, and they are often present at low densities. The simplest way to collect them is to remove a sample of the substrate and search through it by hand, using a sieve, or a series of sieves of various mesh sizes, to remove the larger items. For example, a wooden box about 0.3 m x 0.3 m and 0.25 m deep fitted with a rigid 1 mm^2 mesh is appropriate for sampling sand, while soil can be sieved through larger mesh sizes first to remove bigger pieces of debris. Washing the substrate with plenty of water may cause the lighter animals to float above the inorganic components, at which stage they can be scooped off the surface. The addition of certain salts enhances this process (Southwood & Henderson, 2000).

Hand searching is very time-consuming. However, for moist (rather than wet) substrates, such as many soils and leaf litter, less labour-intensive methods are available. There are several methods using similar principles. The most familiar is the Tüllgren funnel (fig. 13). Here a light shines down on soil held in a coarse sieve positioned above a funnel. The funnel leads to a jar containing a killing and preserving solution such as 70% industrial methylated spirit (IMS*). Animals move down through the soil to escape the light, heat and desiccation and fall through the grid, down the funnel and into the killing fluid. More sophisticated designs seal the light to the top of the funnel and cool the outside of the funnel, increasing the temperature gradient within the apparatus. A simpler extraction method based on the same idea involves placing soil in one side of a tray which is raised slightly at one end. If a light is directed at the raised end, animals tend to crawl out and down towards the lower end. If this contains a killing solution and is shaded, animals will collect there and can be examined later.
Other similar designs include Kempson bowl extractors (fig. 14), which are suitable for small invertebrates such as mites and springtails, and Winkler traps in which the soil is held in small coarse mesh bags suspended in cotton funnels (fig. 15). Winkler traps also work by exploiting soil organisms' aversion to drying out, but they do not use a heat or light source, instead relying solely on the natural drying process. The amount of soil collected can be expressed in terms of

* Industrial methylated spirit (IMS) is cheaper and more widely available than pure alcohol (ethanol). It is diluted 7:3 (IMS:water) for use as a preservative, and known loosely as 'alcohol'.

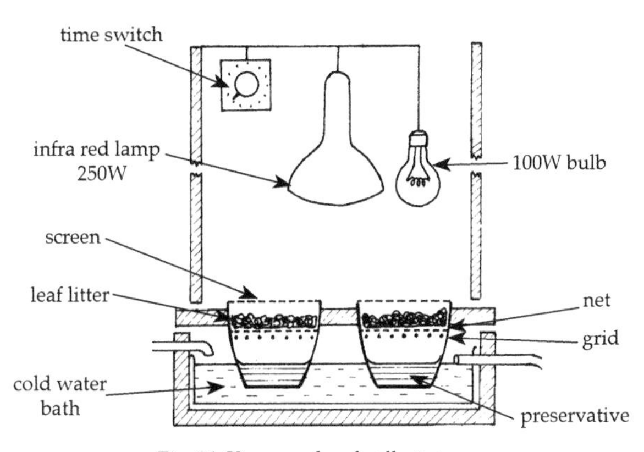

Fig. 14. Kempson bowl collector.

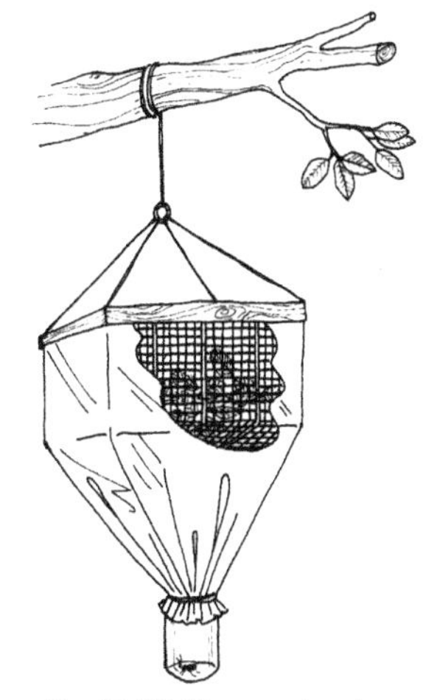

Fig. 15. Winkler trap showing leaf litter held in a bag above a grid through which animals fall to be caught in the preserving fluid beneath.

either the volume or the surface area sampled. Leaf litter is more usually recorded in terms of the surface area removed. These measurements allow counts of invertebrates to be expressed in terms of density of animals per unit volume or unit area.

Some animals can be extracted directly from the ground by adding certain chemicals. Earthworms, for example, come to the surface if a weak detergent solution is poured over the ground. More effective extraction chemicals include weak solutions of potassium permanganate or formalin, but these are toxic both to earthworms and to other organisms. Edwards & Bohlen (1996) give further details of these and more sophisticated extraction methods for earthworms. Other chemicals such as mustard and chilli powder can also be used in suspension as irritants to extract earthworms. East & Knight (1998) compared mustard suspension (at 25 ml of liquid mustard per litre of water) with a variety of strengths of different detergents and found that, not only did mustard collect large numbers of earthworms, but they were more likely to survive than when using detergent. It is important to standardise the collection area, which is usually recorded as the area over which the liquid has been poured and the volume of fluid used.

Some insects live in the soil as larvae and emerge to live on the surface, or on plants, or to fly, as adults. Emergence traps can be used to collect these as they pass from one habitat to another. These traps are usually simple boxes or nets with clear collecting vessels positioned in otherwise opaque traps, exploiting the fact that many animals move towards the light (fig. 16). Southwood & Henderson (2000) give further details.

Methods of sampling are described in more detail by Wheater & Read (1996, NH 22) for soil and leaf litter, and Hayward (1994, NH 21) for marine sands.

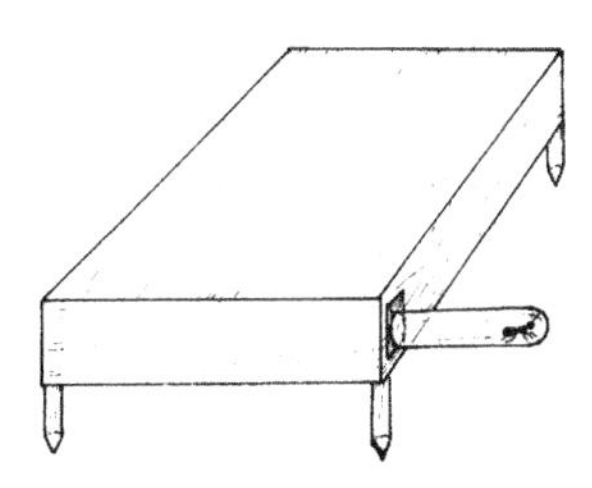

Fig. 16. Emergence trap which is light proof except for the collecting tube where animals gather.

Sampling invertebrates from rocky shores and other hard substrates

The distribution, community structure and densities of sessile or slow-moving organisms such as barnacles, lichens or limpets can be investigated by using quadrats to sample the surface of rocks, walls or even gravestones. If possible these should be recorded *in situ* and left undisturbed, but if necessary they can be removed with care if they do not adhere too strongly. For example, lichens can be freed by sliding a thin-bladed knife between them and the rock, and barnacles by removing a sliver of rock with them on it. As with sampling plants (p. 11), transects may be used to position the quadrats. On rocky shores, if the transects are placed at right angles to the tide line, the zonation of species in relation to tidal level on the seashore can be examined.

Sampling surface active invertebrates

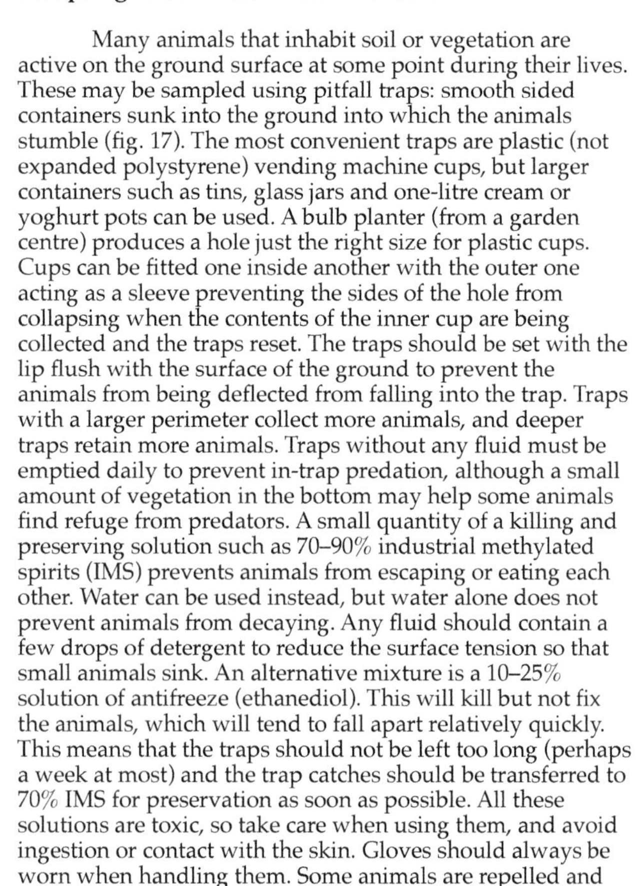

Many animals that inhabit soil or vegetation are active on the ground surface at some point during their lives. These may be sampled using pitfall traps: smooth sided containers sunk into the ground into which the animals stumble (fig. 17). The most convenient traps are plastic (not expanded polystyrene) vending machine cups, but larger containers such as tins, glass jars and one-litre cream or yoghurt pots can be used. A bulb planter (from a garden centre) produces a hole just the right size for plastic cups. Cups can be fitted one inside another with the outer one acting as a sleeve preventing the sides of the hole from collapsing when the contents of the inner cup are being collected and the traps reset. The traps should be set with the lip flush with the surface of the ground to prevent the animals from being deflected from falling into the trap. Traps with a larger perimeter collect more animals, and deeper traps retain more animals. Traps without any fluid must be emptied daily to prevent in-trap predation, although a small amount of vegetation in the bottom may help some animals find refuge from predators. A small quantity of a killing and preserving solution such as 70–90% industrial methylated spirits (IMS) prevents animals from escaping or eating each other. Water can be used instead, but water alone does not prevent animals from decaying. Any fluid should contain a few drops of detergent to reduce the surface tension so that small animals sink. An alternative mixture is a 10–25% solution of antifreeze (ethanediol). This will kill but not fix the animals, which will tend to fall apart relatively quickly. This means that the traps should not be left too long (perhaps a week at most) and the trap catches should be transferred to 70% IMS for preservation as soon as possible. All these solutions are toxic, so take care when using them, and avoid ingestion or contact with the skin. Gloves should always be worn when handling them. Some animals are repelled and

Fig. 17. Pitfall trap.

others attracted by the solutions. Adding baits such as fruit or meat to dry traps may encourage some animals to enter the traps, and this may be useful if large numbers of living specimens are required. There is much scope for research on the selectivity of pitfall traps that contain different preserving fluids or baits.

Fig. 18. Pitfall trap with cover.

Dilution by rain can be avoided by fitting a rain shield (fig. 18): a tile or flat stone can be supported on small stones above the trap or a piece of hardboard can be mounted on 100–150 mm long nails which are pressed into the soil to leave a gap of 50–75 mm between the surface and the shield. Rain shields also help to prevent vertebrates such as shrews from getting into the traps. Surface runoff may also fill traps and cause them to overflow and lose captured animals. This can be avoided by setting traps away from areas likely to flood, and by perforating the traps about 10 mm from the top to allow excess water to drain away.

Fig. 19. Barrier between two pitfall traps.

The perimeter of the trap can be increased by using either larger diameter containers or long channels, such as guttering, although these are more difficult to sink into the ground. Alternatively, barriers placed between pairs of traps increase the effective perimeter of the traps, because many surface-active invertebrates follow edges. These barriers are usually made of thin (about 3 mm thick) pieces of wood, plastic or metal about 150 mm high supported by sinking them about 25 mm into the ground (fig. 19). This idea can be extended to examine movements into and out of an area, around, for example, a log, or a stone, or a particular plant (fig. 20). If the area under investigation is surrounded by a double 'fence' of barriers, each with a pitfall trap at each corner, the outer set will collect animals moving in, while the inner set will catch those moving out.

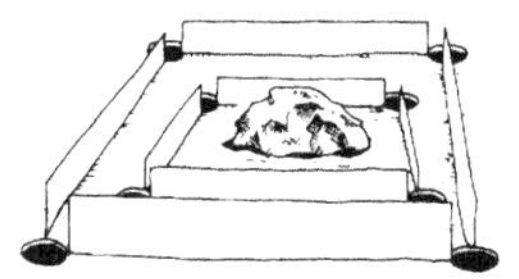

Fig. 20. Barriers surrounding stone.

Wheater & Read (1996, NH 22) and Southwood & Henderson (2000) give more information on sampling surface active animals.

Sampling invertebrates living on and in plants

Plants provide a wide range of habitats for many invertebrate species, including predators such as spiders, ladybirds and rove beetles as well as insects that feed directly on the plant tissue itself such as greenflies, thrips, caterpillars and weevils.

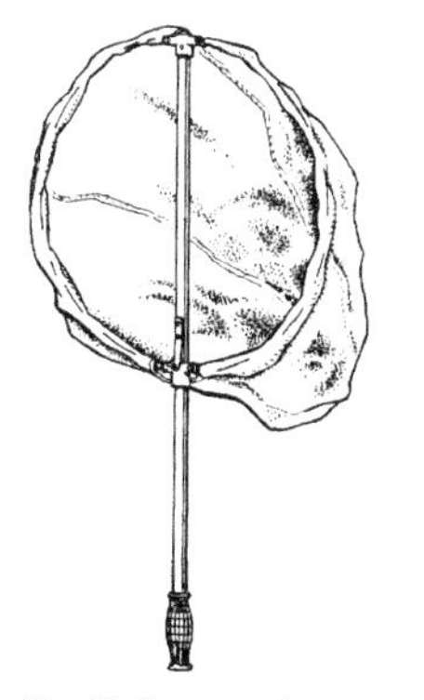

Fig. 21. Sweep net.

Plants can simply be searched visually and the animals picked off by hand. However, this is time-consuming, can disturb the animals causing mobile species to flee, and can be difficult (and painful!) if plants like stinging nettles and brambles are involved. For vegetation such as fungi and dead wood there may be no alternative. On other vegetation, a sweep net (a tough net on a rigid frame, fig. 21) can be used. The net is swung forcefully through the vegetation dislodging the animals and catching them in the net. This works very well with flexible, erect plants such as nettles, docks, or grasses, but not on lower-

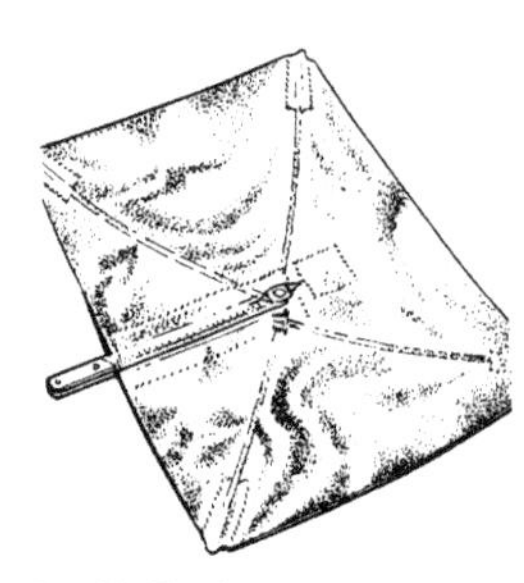

Fig. 22. Beating tray.

growing or woody species such as cabbages or trees. Sweep netting can damage the habitat, so take care to sweep only part of a site. On trees and shrubs beating can be useful. Sheets of cloth or polythene (sometimes held on a frame, fig. 22) are positioned below the tree or shrub and the plant is then struck firmly several times with a stick to dislodge animals onto the sheets for collection. Here they must be collected quickly or some will escape. Others may avoid capture by hanging on securely.

Small animals can be picked up with fine forceps. Rigid forceps are likely to damage fragile specimens; soft stork-billed forceps are more useful. Very small animals can be sucked up using a pooter. Pooters are available commercially (suppliers' addresses pp. 99–100), but they are easily made from a small plastic container sealed with a lid or a bung with two rubber tubes leading from it (fig. 23). The collector sucks on one tube whilst holding the end of the other one close to the animal to be collected and the animal is then sucked into the collecting vessel. The inner end of the sucking tube is covered with gauze to avoid getting a mouthful of animals. The best designs have the tube being sucked wider than the one held over the animal, creating more suction. A much simpler pooter can be made from two lengths of polythene tubing differing in diameter so that the narrower one fits snugly inside the wider tubing. They are joined end to end by pushing one into the other, trapping a gauze barrier between them. The animals, once caught, can then be blown out into a collecting tube.

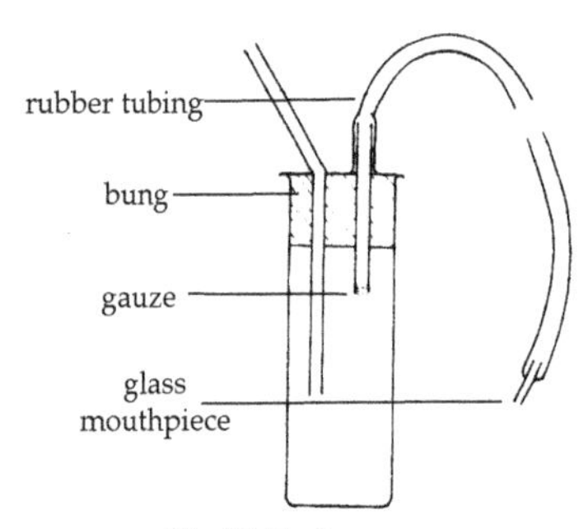

Fig. 23. Pooter.

Suction samplers can be used to extract animals from the surface of low growing plants. However, traditional suction samplers are heavy and expensive. A lighter and cheaper alternative can be made from a petrol driven garden vacuum cleaner modified by placing a net over the inside of the nozzle so that insects are caught in the net when sucked up (fig. 24). Even cheaper is a similarly modified battery-operated car vacuum cleaner, although since these have relatively poor suction pressures, they are often less effective. The net will also catch leaf litter and debris, which can be separated out on a white sheet. Large quantities of debris make sorting difficult. On loose substrates sharp gravel may damage the netting. To maximise the suction for a given power of sampler, the aperture of the nozzle should not be too wide (Bell and others, 2002). A couple of seconds per suck is long enough for many animals. The total area sampled should be calculated by counting the number of sucks and multiplying it by the surface area of the nozzle.

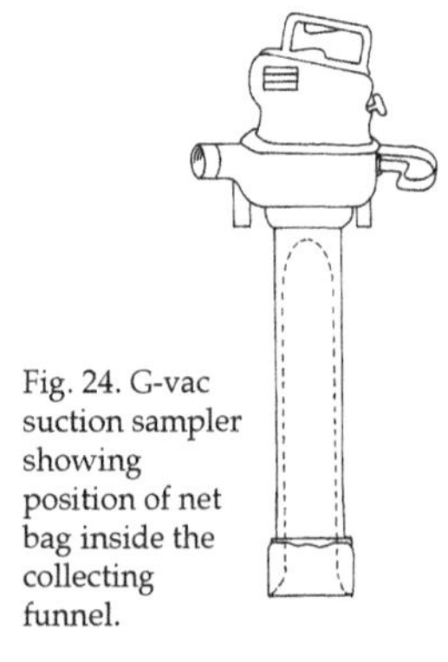

Fig. 24. G-vac suction sampler showing position of net bag inside the collecting funnel.

Animals living inside plants can be extracted by dissection, or by caging the relevant part of the plant and capturing the animal on emergence. For example, some moth larvae produce mines in leaves. They can either be dissected out, or left *in situ* with the leaves enclosed in a muslin sleeve until the adult moths emerge. More information on leaf miners on several different plant species is given by Davis (1991, NH 1), Redfern (1995, NH 4), Kirk (1992, NH 18),

Salt & Whittaker (1998, NH 26), and Leather & Bland (1999, NH 27). Gall-forming insects can similarly be reared by enclosing galls containing fully-grown larvae in bags made of net curtain material or nylon tights (Redfern & Askew, 1992, NH 17). Dense plants such as cabbages, or the larger seaweeds, can be enclosed whole in netting or a polythene bag for removal to the laboratory, where careful inspection will reveal small animals that may not be easy to find in the field. If the bag containing the plant is left in a refrigerator for a short time, the animals slow down, making sorting easier. Very small plants, such as mosses, are taken to a microscope for the identification of microscopic animals that inhabit them (Hingley, 1993, NH 20). Remember that whole plants should not be removed in this way without the land owner's permission. Animals that live under the bark of trees or decaying wood can be sampled by peeling back the bark and picking them up with a pooter or stork-billed forceps. Take care to replace bark on dead wood, and to avoid damaging bark on living trees.

Sampling aerial insects

Catching aerial insects can be hard work. Hand nets such as butterfly nets can be used to sample flying insects. Practice is required to be sure that the technique is used consistently, especially with fast-flying and evasive species such as dragonflies.

More passive techniques involve setting traps that either attract the animals or intercept them during normal behaviour. Water traps attract a wide range of flying insects. These can be as simple as a series of coloured bowls (from dessert bowls up to washing-up bowls in size) containing water with a few drops of detergent to prevent insects from escaping from the water surface. If comparisons are being made, for example, between habitat types, then the size, shape and colour of the bowls must be uniform. Colour is particularly important since different ecological groups seem to be attracted to different colours: grass-feeding species (such as some thrips and greenflies) appear to have few colour preferences, whereas non-grass foliage feeders (and their associated predators and parasites) seem to prefer yellow traps. It is also possible that insect pollinators such as bees and hoverflies are more likely to be caught in orange, yellow and white bowls whereas horseflies may prefer darker colours (brown, dark blue and purple). There is much scope for further research to identify the most efficient colours for catching particular ecological or taxonomic groups of insects in different habitats (Kirk, 1984).

Colour preferences also influence the efficacy of sticky traps, which are available from many horticultural suppliers. These are plastic sheets covered with a non-setting glue from which animals cannot free themselves once attached. Many sticky traps are bright yellow, and would attract leaf-feeding species and insects that visit flowers.

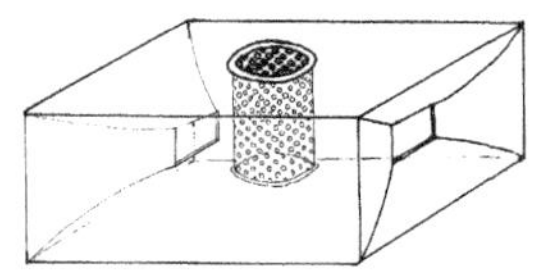

Fig. 25. Assembly trap showing mesh container to hold virgin females and slanted ends to trap males investigating pheromone plumes.

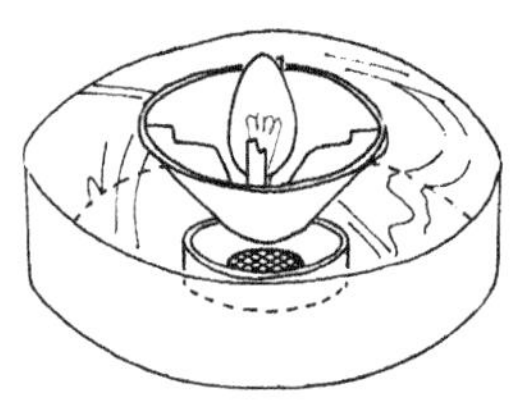

Fig. 26. Robinson moth trap.

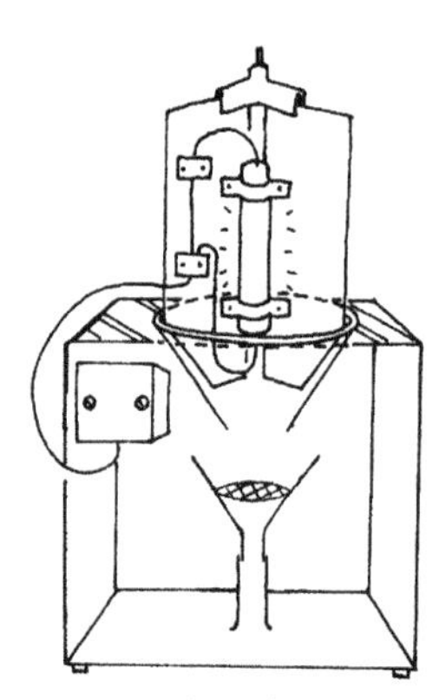

Fig. 27. Heath moth trap.

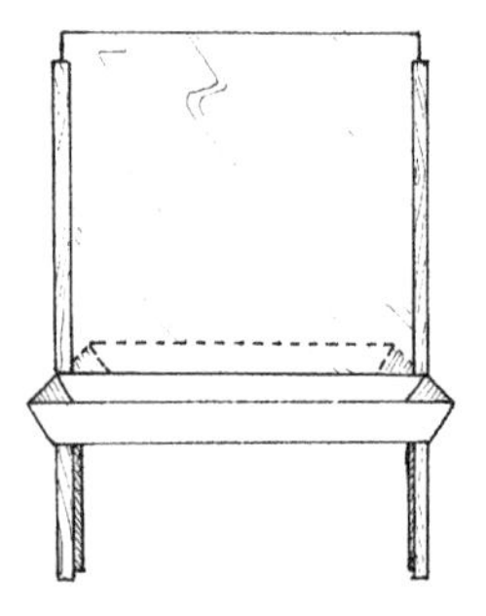

Fig. 28. Window trap.

These traps can be made up using non-setting glue such as Tanglefoot (from the Tanglefoot Company, address p. 100), or tree banding grease (from horticultural suppliers) or petroleum jelly. Other sticky materials may act as baits, as well as preventing the insects from leaving the trap. For example, a thick mixture of molasses, brown sugar and stout can be painted on trees to attract and hold moths feeding at night. The catch of a sticky trap may include not only animals attracted to them (for example by molasses or the yellow colour), but also more passively-moving animals in the airstream. Some of these may not even be flying. Some small beetles are blown by the wind, and many small spiders use a length of silk as a sail to float along on air currents.

Another way of attracting some insects to traps is by exploiting the ways in which one sex attracts the other for mating. In some moths, such as scarlet tiger moths, virgin females attract males by releasing a chemical signal, or pheromone. Female scarlet tiger moths are often found in undergrowth. If one is held in a cage, males assemble and can be caught in a surrounding trap (fig. 25). Similarly, female crickets can be collected when they are attracted towards a singing male cricket in a cage.

Several nocturnal insects (especially moths, but also some beetles such as weevils) can be collected using light traps. The traps disrupt normal flight behaviour and work best in the absence of moonlight, under closed canopies, or when the sky is overcast. There are several types of light trap currently available. Most light traps catch the animals alive, although some have facilities for killing agents to be added (Southwood & Henderson, 2000). The most efficient at both capture and retention are large traps that use mercury vapour lights such as the Robinson trap (fig. 26). Smaller portable traps, such as the Heath trap (see fig. 27), often use fluorescent tubes powered by motorcycle batteries. These do not attract as many insects as mercury vapour traps, but are easier to transport. A portable trap called a Skinner trap that uses Robinson trap lights is available from Watkins & Doncaster (address p. 100). Fry & Waring (1996) give details of the construction and operation of the popular types of moth traps, and several suppliers stock a range of types of light traps (addresses pp. 99–100).

Various static traps can be used to intercept insects in flight. Window traps (made of clear plastic, fig. 28) and interception traps (made of fine netting, fig. 29, and available from Marris House Nets, address p. 99) can be set across likely flight paths. Flying insects, and other animals blown on the wind, are collected in trays at the base of the trap.

Static suction samplers catch insects as they fly past, but are expensive. A Malaise trap can also be used to trap flying insects (suppliers' addresses pp. 99–100). Here a vertical panel of netting intercepts insects in flight and a sloping tent-like roof of netting guides them, as they crawl upwards, into a collecting jar which can contain a killing solution, or can be fitted with a one-way valve if living

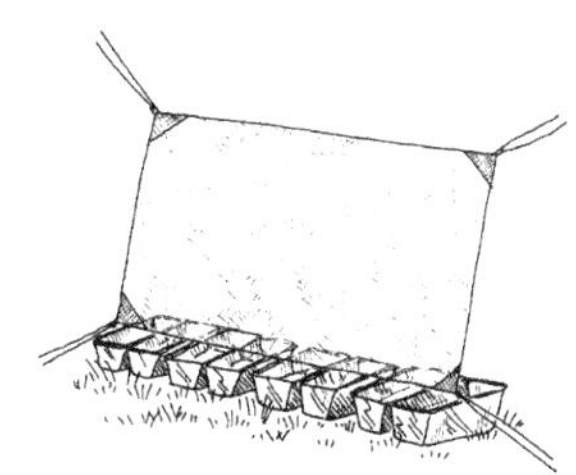
Fig. 29. Interception trap.

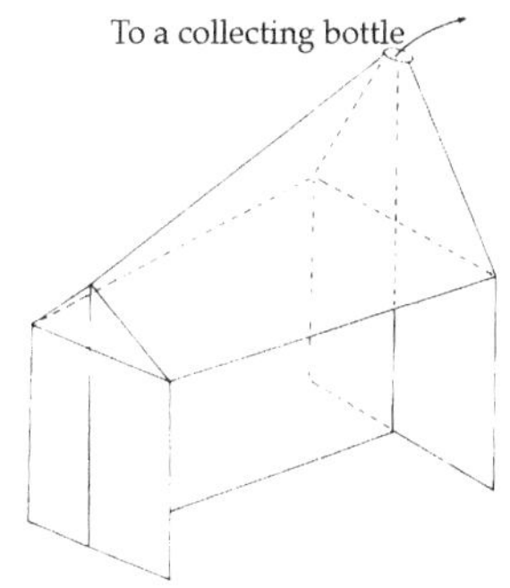

Fig. 30. Malaise trap.

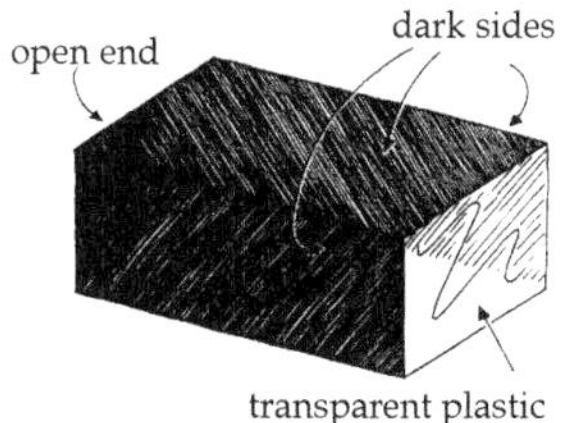

Fig. 31. Herting trap.

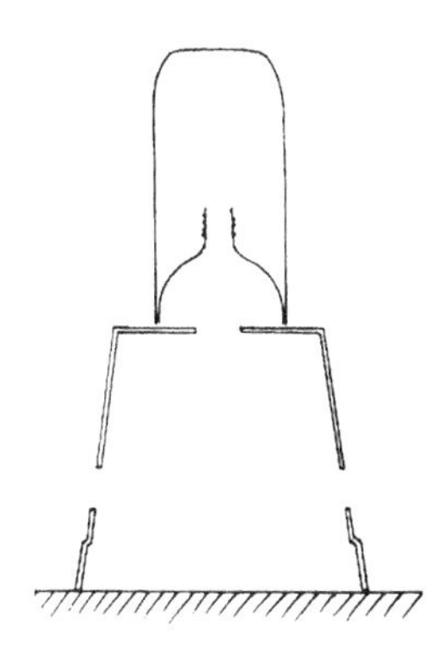
Fig. 32. Flower pot trap based on Irwin trap.

specimens are required (fig. 30). Strongly flying insects can be collected in a Herting trap (fig. 31). This is a tube of dark material with one open end for entry and clear plastic film or fine netting over the other end. Insects such as blowflies will fly along the tube towards the netting and tend to stay there. Erzinçlioğlu (1996, NH 23) gives further details. Animals such as blowflies that are attracted to carrion can be caught in an Irwin trap baited with a dead mammal or a piece of meat. An Irwin trap is simple to make by cutting the top end off a plastic drinks bottle and inserting it neck first into the remainder of the bottle. Insects are attracted to the bait inside and most cannot escape. An Irwin trap is more efficient if placed vertically, mouth down, on top of an inverted plastic plant pot which has had two 1-cm wide slits cut into its sides near the rim and a 3-cm diameter hole cut in its base (fig. 32). Erzinçlioğlu (1996, NH 23) gives details.

Marking invertebrates

It is often necessary to mark animals, for example to discover whether any animals caught are resident or have moved within the environment, to estimate population size (p. 51), or to tell individuals apart if kept in groups in a container or cage. If whole categories of animals are to be marked alike, such as all beetles caught in a given area or during a given sampling period, a simple blob of paint or similar material can be put on each animal caught. The best type of marking material depends on the species to be marked and the position of the mark. Paints can be used on hard cuticles, and felt-tip pens on wings. For nocturnal animals, fluorescent powders that glow under ultraviolet light can be useful (BioQuip, address p. 99). Torches with ultraviolet light bulbs or tubes are available from many suppliers (addresses pp. 99–100).

Sometimes animals need to be marked so that they can be recognised individually, for example to determine how frequently they move between plants or other habitat features such as logs and stones. Individual marking also helps in some methods of estimating population size. Snails are easily numbered using waterproof paint, or by gluing numbered tags onto their shells or using a small craft drill to engrave the shells. Animals such as beetles can be marked using a coding system in which dots of different colours and positions on the back of the animal give a large number of different combinations. The system used in fig. 33 is an example. Other coding methods are described by Davis (1991, NH 1) and Majerus & Kearns (1989, NH 10). A flexible, waterproof, fast-drying paint such as cellulose dope or acrylic paint is suitable. Flying insects such as butterflies and moths can be marked on the wings with a fine felt-tip pen or quick-drying paint, using a system of dots or, on species with large pale areas of wing on which to write, using numbers.

However careful you are, capture, handling and marking are likely to influence the animal's behaviour and

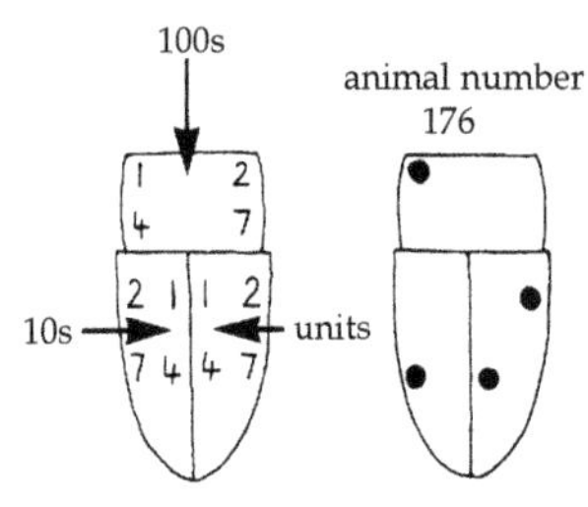

Fig. 33. Marking code.

its chances of survival. For example, damaged individuals may not mix freely with the rest of the population, and conspicuous coloured marks may reduce or enhance predation. This disturbance cannot be entirely eliminated, but it can be minimised if you are careful and quick. Marks should therefore be applied quickly and neatly to minimise disturbance of the animal's normal behaviour. Animals can be restrained in nets, or slowed down by short periods in a fridge or insulated cool box containing ice packs, or if necessary knocked out using carbon dioxide. The type of wine bottle opener that uses carbon dioxide to blow the cork out is useful for injecting small quantities of the gas into tubes, especially in the field. A similar sparklet-operated carbon dioxide dispenser is available from Edme Ltd (address p. 99). Similarly, freshwater invertebrates may be immobilised by adding cheap sparkling (carbonated) bottled water to the water in which they are held. Invertebrates tend to recover quite quickly from these treatments and appear to suffer no ill effects. The use of other chemical agents should be avoided where possible since they are likely to affect the animals' behaviour.

Keeping live invertebrates

Sometimes it is necessary to keep live invertebrates, either to allow observation of behaviours such as food preferences, defensive postures or mating, or to obtain adults for taxonomic verification of hard-to-determine life stages such as eggs and larvae.

Some animals do well in captivity. Decomposers such as woodlice, millipedes and springtails can be kept in a cool place in a plastic lunch box with a good thick layer of leaf mould. This should be moist, but not too wet. The leaf mould needs to be topped up with fresh fallen leaves occasionally, and vegetables such as slices of potato or cucumber can also be added as food. Some of the smaller predators can also be kept in this way, eating tiny soil invertebrates. Larger predators, such as large ground beetles, large spiders and centipedes, need to be fed more actively and many will only take live prey. Substitutes for naturally available food are aphids, vestigial-wing *Drosophila* or the maggots available in most fishing-tackle shops. Those accepting dead food may take small amounts of minced meat or dried tubifex (available from tropical fish shops). Many plant-feeding larvae can be raised by keeping them in cages on the host plants on which they were found, with dead or dying plants replaced regularly (fig. 34). Flighted adults may require larger cages and a different food type. Adult butterflies and moths will survive longer if fed on sugar solution placed on cotton wool, or added to fresh or artificial flowers. Aquatic animals need to be kept in well-aerated water or water that is changed regularly, and given appropriate food. Chlorinated tap water should be allowed to stand for several hours before

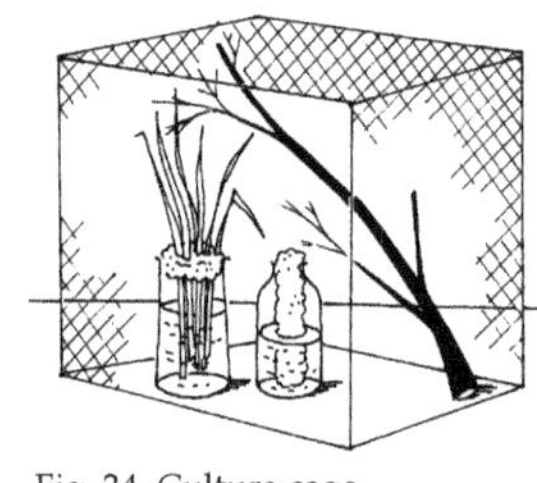

Fig. 34. Culture cage.

use. Survival will be enhanced if the animals are kept cooler than the ambient air temperature.

Keeping animals in the fridge for several days may increase their life span by slowing development. This is useful if they are needed for experiments. Animals that normally undergo a seasonally induced diapause (period of arrested development) may be kept active by keeping them in a light cycle that simulates summer (that is has more daylight hours than dark ones). Ensure that closed containers have sufficient air holes, and avoid the build up of condensation. Open containers are free from such problems, but may require a band of grease near the top to prevent animals from crawling out. Tree banding grease is available from garden centres.

Lutz and others (1959) give information on rearing a wide range of invertebrate animals, and details for particular groups are given by Brown (1990, NH 2) for grasshoppers; Gilbert (1993, NH 5) for hoverflies; Forsythe (2000, NH 8) for ground beetles; Majerus & Kearns (1989, NH 10) for ladybirds; Rotheray (1989, NH 11) for aphids and their predators; Harker (1989, NH 13) for mayflies; Snow (1990, NH 14) for mosquitoes; Morris (1991, NH 16) for weevils; Kirk (1992, NH 18) for the larvae of butterflies and moths; Erzinçlioğlu (1996, NH 23) for blowflies; Skinner & Allen (1996, NH 24) for ants; and Kirk (1996, NH 25) for thrips.

Killing and preserving invertebrates

Do not kill animals without first considering whether it is necessary and what the alternatives are. For many groups of invertebrates, critical identification unfortunately requires close examination of dead and preserved specimens. Once the commonest species have been learnt in this way it is often possible for some identification to be done in the field without killing the animals.

Most soft-bodied animals can be both killed and preserved by immersion in 70% industrial methylated spirit (IMS). The sale of this is controlled by law in Britain and it may be difficult to get hold of unless you have links with a school, museum or university. Surgical spirit (from chemists) or methylated spirit (from hardware stores) are possible alternatives. However, both may cause distortion of the specimen and the former tends to deposit an oily layer on specimens, while the latter contains a dye. Alcohol evaporates quickly and some animals, such as spiders, are virtually impossible to identify once they have dried up. A drop of glycerol added to the alcohol slows evaporation (and the stiffening of specimens). Specimens preserved in 70% alcohol should be stored in vials that can then be submerged in alcohol in glass bottles or jars with tight-fitting lids. This double protection reduces the likelihood of the specimens drying out. Each vial, bottle and jar should be well labelled. Ballpoint ink is soluble in alcohol, so use pencil or indelible

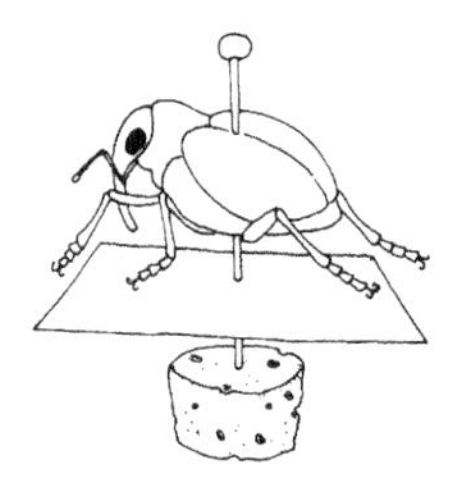
Fig. 35. Pinned insect.

ink for labels. Soft paper breaks down in time, so for long-term preservation museums use special goatskin parchment paper for labels (suppliers' addresses pp. 99–100). Some soft-bodied animals need to be fixed before preservation to reduce distortion in storage. Lincoln & Sheals (1979) give more information on preservation in fluid.

Small insects (such as gall midges) and other arthropods (such as mites) will eventually need to be mounted on slides for examination with a compound microscope. Different groups are mounted in different media. The relevant identification guide (Chapter 3) will indicate the preferred mounting medium and method of preparation for the group in question.

Fig. 36. Insect pinned on foam.

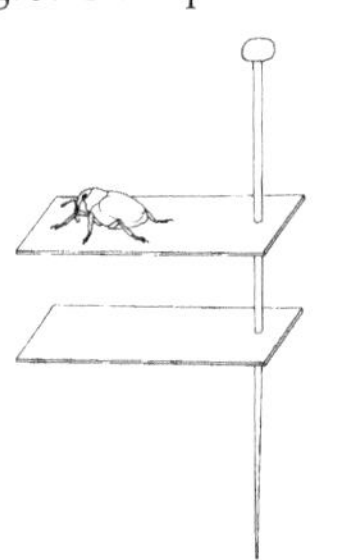
Fig. 37. Carded insect.

Hard-bodied insects, and those with obvious wings, such as beetles and butterflies, may be killed in a closed jar or vial in the bottom of which is a layer of set plaster of Paris, or some blotting or filter paper, moistened (but not wetted) with a few drops of ethyl acetate. Small insects are killed after about 30 min exposure to ethyl acetate vapour, but robust beetles and bumblebees need much longer. Alternatively, insects can be killed in a deep freeze. They can then be mounted. Robust specimens can be mounted directly on an entomological pin (suppliers' addresses pp. 99–100) pushed through the thorax (or, for beetles, through the right wing case and the abdomen) (fig. 35). A smaller specimen is mounted on a very small headless entomological pin. This in turn is mounted on one end of piece of polyporus or polyethylene foam (suppliers' addresses pp. 99–100) the shape and size of half a matchstick, the other end of which is mounted on a standard entomological pin (fig. 36). Small beetles are sometimes mounted by gluing their feet and antennae to small rectangles of thin card or clear acetate sheet with water soluble glue and mounting the card on a standard entomological pin (fig. 37). An alternative method is to glue the animal to a triangle ('point') of card which is then mounted on a pin (fig. 38). In each case, labels should be put on the pin below the specimen, giving the date and place of capture, together with other information, such as who caught it. The specific name and the name of the person who identified it should be on a separate label on the same pin. Large-winged insects like butterflies are sometimes set by being impaled on a pin through the thorax and being pinned out on a cork setting board (fig. 39; suppliers' addresses pp. 99–100) or a sheet of polyethylene foam, with the wings held in position by pins or by paper strips pinned to the board or foam. They are allowed to dry in this position before being transferred to a storage box or cabinet. Animals that are stored dry are vulnerable to attack by small arthropods and fungi. A range of chemicals, many of which are toxic, is available to protect them, but the most important protection is to store them in a box with a tight-fitting lid. Wooden insect boxes lined with cork or polyethylene foam are expensive (suppliers' addresses pp. 99–100), but a clear sandwich box with a tight-fitting lid and a floor of polyethylene foam makes a satisfactory alternative.

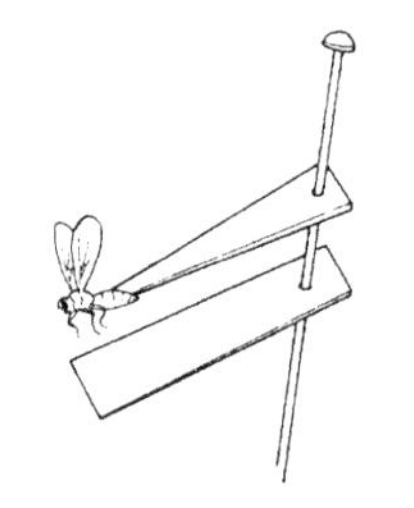
Fig. 38. Insect mounted on a card point.

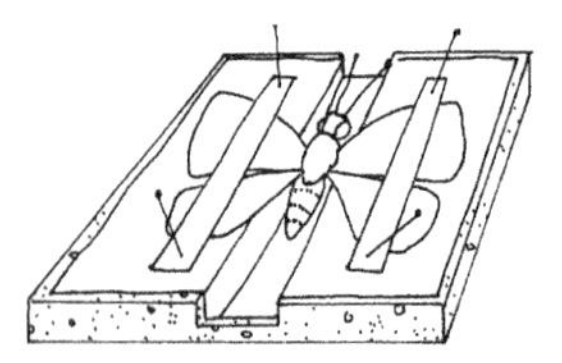
Fig. 39. Set butterfly.

PLATE 1

Marine invertebrates

1. Sea slug
2. Razor shell
3. Crab
4. Sea urchin
5. Starfish
6. Brittlestar

1

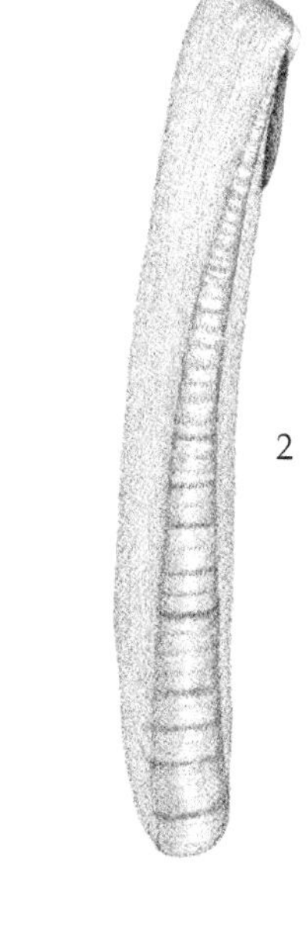

2

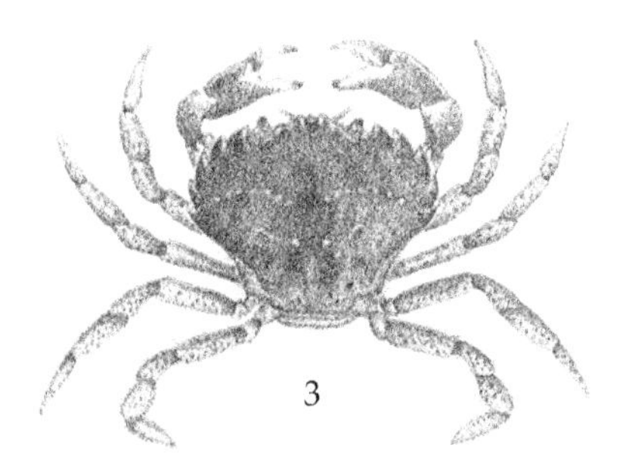

3

4

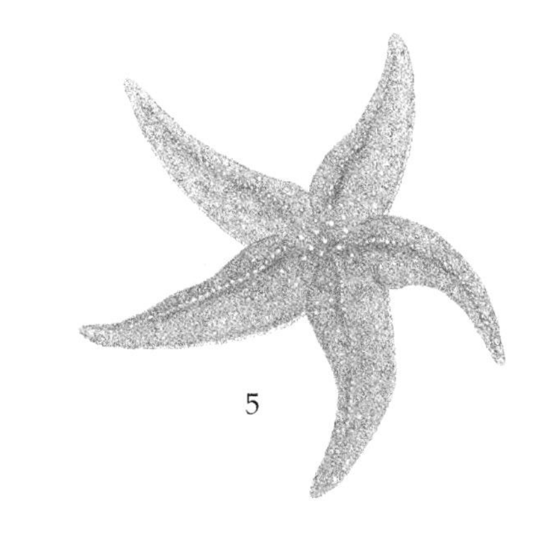

5

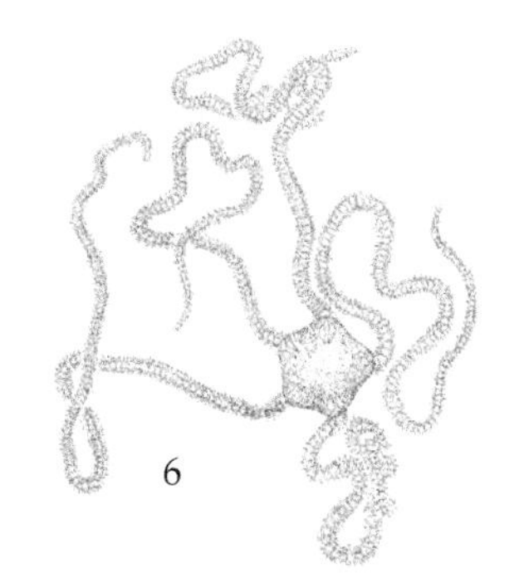

6

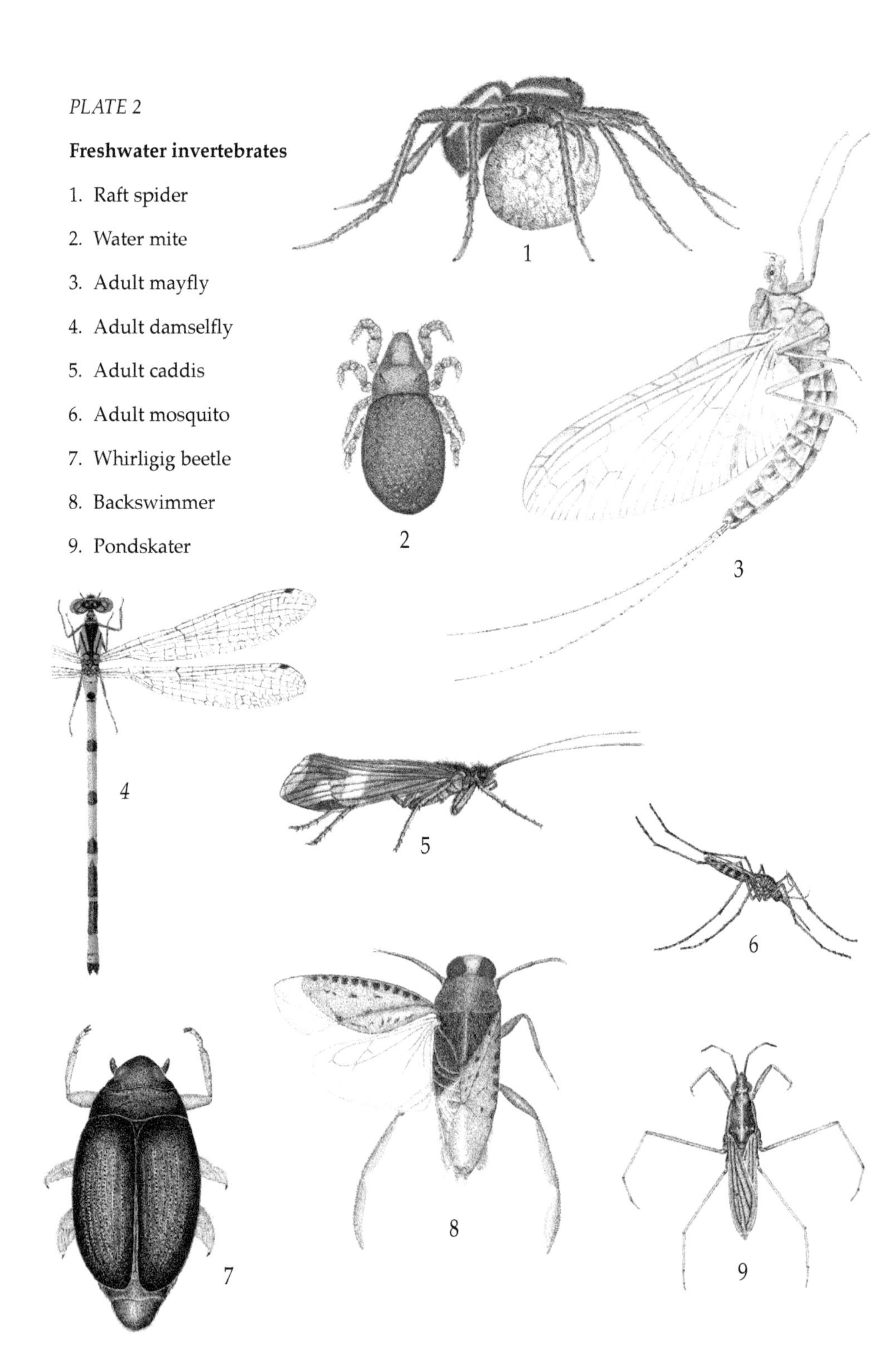

PLATE 2

Freshwater invertebrates

1. Raft spider
2. Water mite
3. Adult mayfly
4. Adult damselfly
5. Adult caddis
6. Adult mosquito
7. Whirligig beetle
8. Backswimmer
9. Pondskater

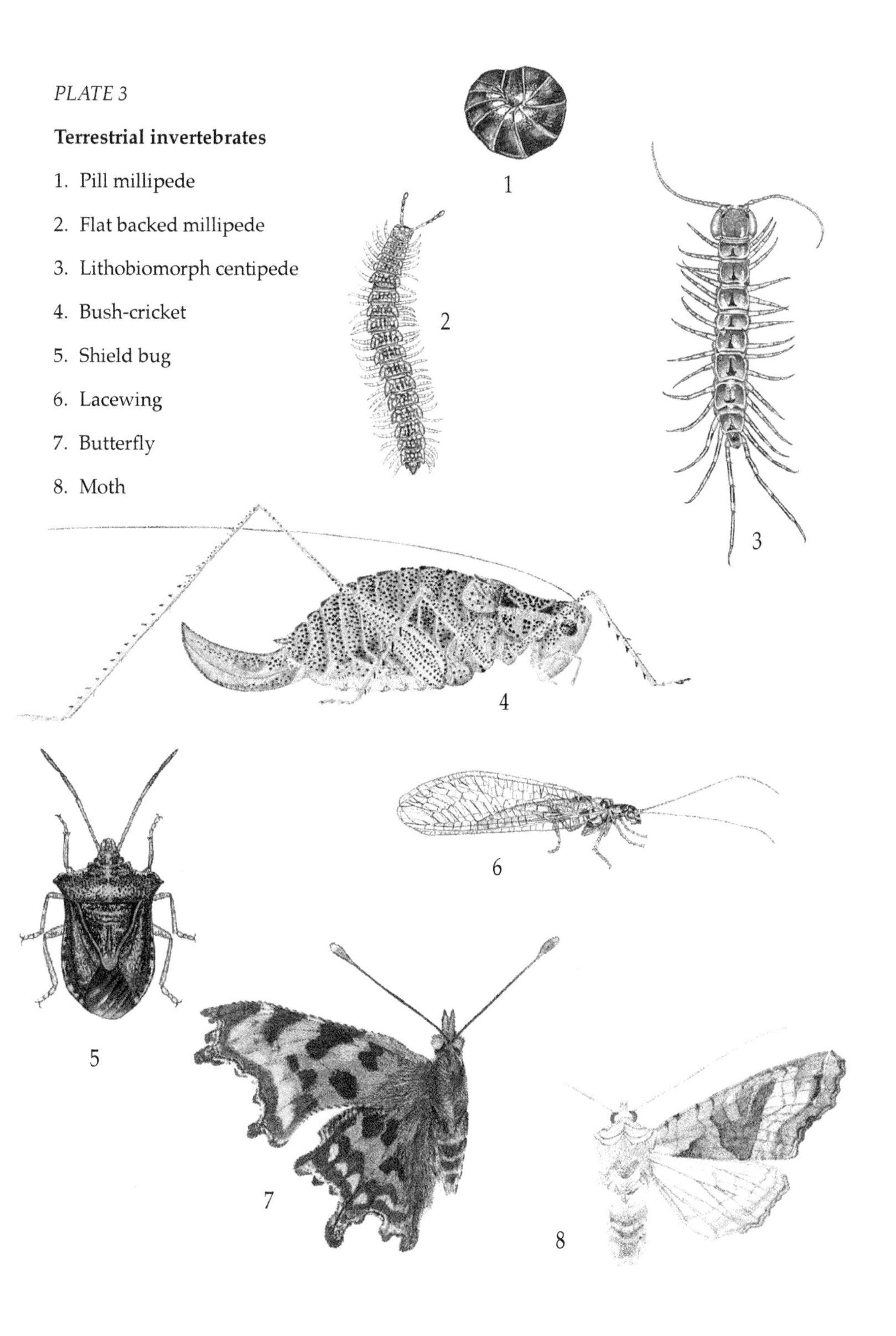

PLATE 3

Terrestrial invertebrates

1. Pill millipede
2. Flat backed millipede
3. Lithobiomorph centipede
4. Bush-cricket
5. Shield bug
6. Lacewing
7. Butterfly
8. Moth

PLATE 4

Terrestrial invertebrates

1. Hoverfly
2. Blowfly
3. Bumblebee
4. Wasp
5. Weevil
6. Leaf beetle
7. Ladybird

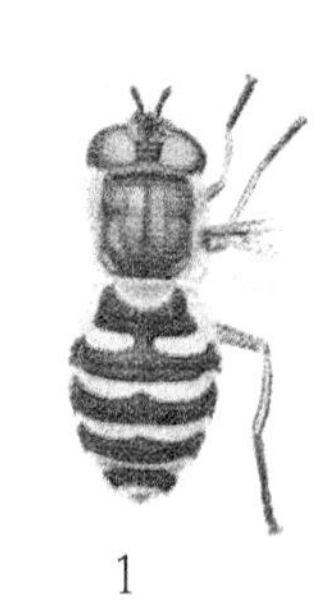

1

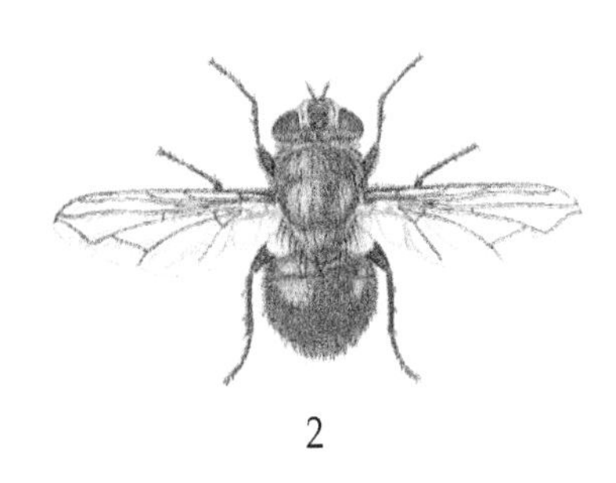

2

3

4

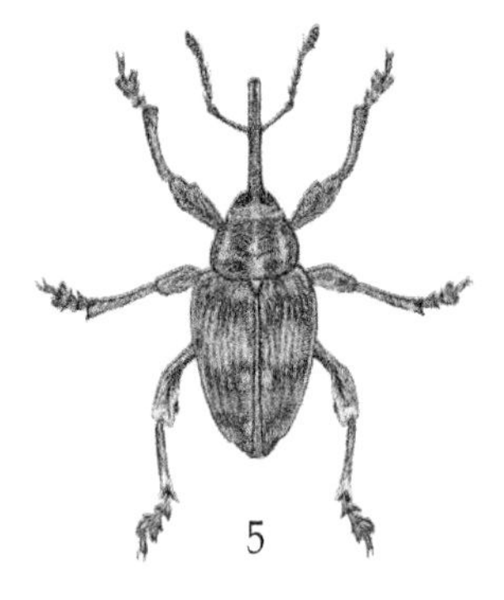

5

6

7

3 Identifying invertebrates

Although it is possible to study animals that you have not identified, this poses several problems. Similar-looking animals may belong to different species with very different behaviour and ecology, but could be combined inadvertently in a study unless they were correctly identified. If you do not know the species name you cannot benefit from published information about the animal , and your findings are of less value to other people. Identification to species level is the most useful, because a species is usually considered as a real biological entity, whereas higher taxa, such as families, vary with the opinion of the taxonomist who created them: one taxonomist might classify a group into a few families, each with many species, whereas another might prefer to establish more numerous, smaller families.

Identification usually requires a x10 or x20 hand lens or preferably a binocular dissecting microscope with a good incident light source (Philip Harris or Watkins & Doncaster, addresses pp. 99–100). An angle-poise lamp is often useful to provide a second light source on the other side. When observing animals preserved in alcohol under a microscope, keep them immersed in fluid to avoid distortion of the image by the meniscus. If available, tiny glass beads are useful for positioning the specimens. Well-washed river sand is a suitable alternative. Fill a solid watch glass or small petri dish approximately one third full of beads, and add 70% industrial methylated spirit (IMS) and then the animal to be observed. Close examination of preserved specimens may force you to breathe the fumes of the alcohol or other preservative. This can be avoided by submerging the specimen in water for examination, but it should be returned to preserving fluid immediately afterwards.

To measure the animal, it is usually sufficient to use a sheet of graph paper under the petri dish and estimate lengths using the squares. More accurate measurements of small animals, or parts of animals, require an eyepiece micrometer (Philip Harris, address p. 99). Guthrie (1989, NH 12) and Hingley (1993, NH 20) give further details.

It is often useful to observe animals alive with a hand lens or a microscope, but to do so it may be necessary to restrain them. This can be done by placing them in a small polythene bag and using a ring of metal to isolate them in a 'bubble' of polythene (fig. 40). Cooling in a refrigerator for five or ten minutes will slow animals down for a short time. Although this is rarely long enough to allow detailed examination, it may help if the identification of a familiar animal is being confirmed. Freshwater invertebrates can be slowed down by adding cheap sparkling (carbonated) bottled water to the water in which they are held.

Fig. 40. Restrained animal.

A major requirement for accurate identification is an appropriate key, handbook or field guide. Some taxonomic groups (such as many plants and some invertebrates including butterflies and dragonflies) can be identified in the field using field guides, handbooks or identification keys. Field guides based on pictures can be useful for larger organisms, such as most butterflies, that differ in obvious characteristics, but they cannot be used for critical identification unless all species in the group are shown, or lists of diagnostic characters are given (as in Gilbert, 1993, NH 5). Naturalists' Handbooks use readily-observed characters where possible, and most of the titles cover all the British species in a group. Those that cover only the commoner species take account of rarer and more difficult species in the keys and direct the user to more specialised texts at appropriate points. AIDGAP keys cover small groups of organisms in depth (for example Hopkin, 1991, takes woodlice to species and Plant, 1997, does the same for lacewings), or larger groups to higher taxonomic levels (for example Tilling, 1987, for major groups of terrestrial invertebrates, Croft, 1986, for freshwater invertebrates and Unwin, 1988, for families of beetles). These keys can be used without specialist knowledge, and some are appropriate for use in the field. For species separated on very small characteristics, specialist works and detailed microscopic examination may be needed. Specialist identification keys often require substantial expertise. Some species-rich groups of small invertebrates can be particularly challenging to identify.

Three professional bodies produce keys for the identification of British invertebrates. The Freshwater Biological Association (address p. 98) concentrates on the identification of aquatic invertebrates, but also provides keys to the flighted adults of some insect groups whose larvae are aquatic (for example Hynes, 1977, for stoneflies). The Royal Entomological Society (previously known as the Royal Entomological Society of London) produces a series of *Handbooks for the Identification of British Insects* (address p. 99). The series does not yet cover all groups of British insects, and some titles are now out of print. However, new titles and revisions of older titles are constantly being added. The Linnean Society of London (address p. 98) produces the *Synopses of the British Fauna* series. There are two series, old and new, the latter being a progressive replacement of the former. This series covers marine, freshwater and terrestrial invertebrates and provides specialist identification to species for defined groups such as land snails (Cameron & Redfern, 1976) and earthworms (Sims & Gerard, 1999).

Key works for the identification of some British invertebrates (insects and arachnids) are listed by Barnard (1999). A shorter review of key works for insects can be found in Corke (1999). The following section summarises some of the specialist and non-specialist guides to the identification of common invertebrate animals in marine, freshwater and terrestrial habitats. Be aware that there is some overlap between these habitat types, for example some terrestrial flying insects have aquatic larvae, and some estuarine waters have species from both marine and freshwater environments, as well as animals that specialise in

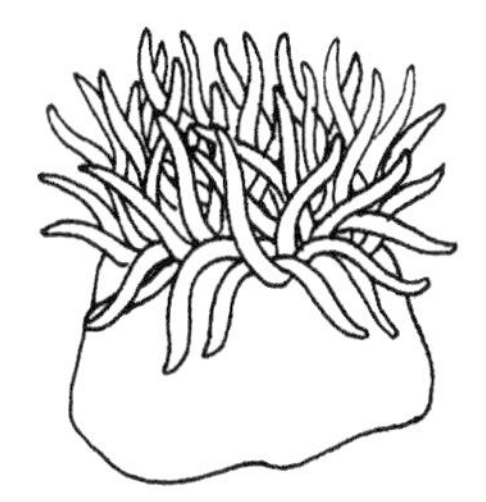

Fig. 41. Sea anemone.

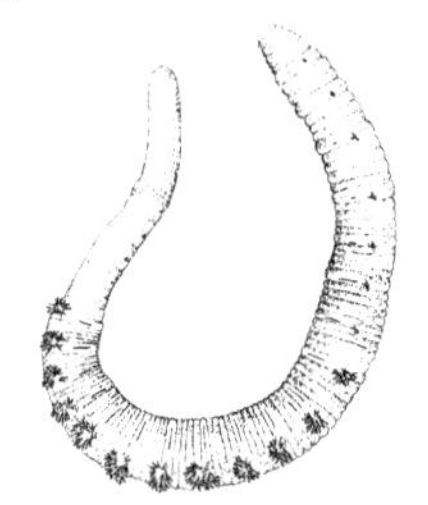

Fig. 42. Lugworm.

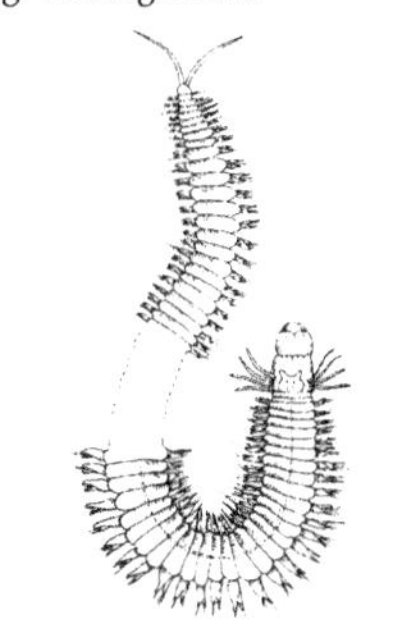

Fig. 43. Ragworm.

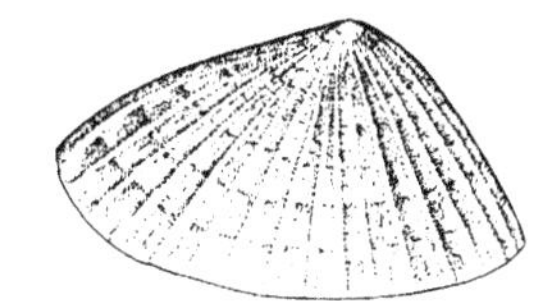

Fig. 44. Limpet.

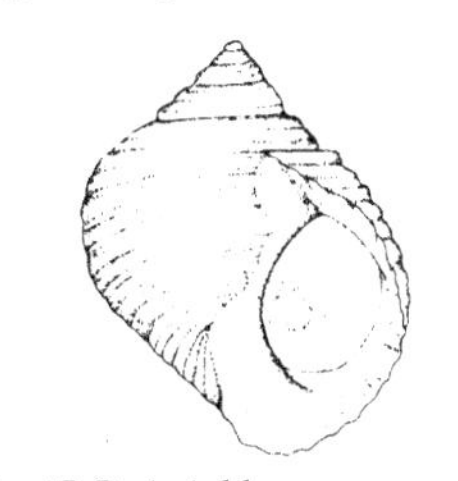

Fig. 45. Periwinkle.

brackish waters. Therefore you should examine more than one section where relevant. It is also important to recognise that guides such as those mentioned do go out of date. The taxonomy of animal groups is a dynamic science: new species are discovered, old systems are changed, names of animals alter and new animals move into areas from which they have previously been absent. Be aware of the limitations of older keys.

Marine invertebrates

Many marine invertebrate studies take place on the shoreline above or just below the water line. Deeper water animals are very different from those commonly found on the shore. Care should be taken to make sure that specimens are resident and have not simply been washed onto the shore. Hayward & Ryland (1990, 1995) are major works covering the British marine fauna, the latter being a concise version of the former. Three books provide identification keys to a range of marine species: Crothers (1997) covers the major groups, while Hayward (1988, NH 9) and Hayward (1994, NH 21) deal with animals on seaweeds and animals on sandy shores respectively. The Linnean Society of London (address p. 98) produces keys to some groups, and there are several field guides that cover the marine environment, for example, Hayward and others (1996), Fish & Fish (1996), and Campbell (1984). Brackish water animals are also covered in depth by Barnes (1994), including all those likely to be found in Northwestern Europe. Below, we consider some of the major groups of larger marine invertebrates, and suggest aids for their identification. Unless otherwise stated, these identification guides contain keys to species. Some animals are more amenable than others to investigation if you are relatively inexperienced. For example, lugworms, many molluscs (especially limpets and some periwinkles) and crabs are relatively easy to catch, observe and identify and offer the opportunity for a wide range of investigations.

Sea anemones have thick, upright cylindrical bodies with the central mouth at the top surrounded by whorls of tentacles (fig. 41). They are often found in rock pools and below the water line on rocky shores. Manuel has produced both a colour guide (1983) and a more specialist key (1988).

A variety of worms may be seen in the littoral zone, especially burrowing in sandy shores. Bristle worms (polychaetes) are commonly found and include lugworms (burrowing, firm-bodied, rather leathery, cylindrical worms with small clusters of bristles along the sides: fig. 42) and ragworms (active, slightly flattened worms with many segments, having bristles on well-developed fleshy lobes: fig. 43). Polychaetes are covered best by Nelson-Smith & Knight-Jones (1990).

Molluscs are found especially on rocky shores and the snail-like prosobranch gastropods are most common. Molluscs include limpets (fig. 44), periwinkles (fig. 45),

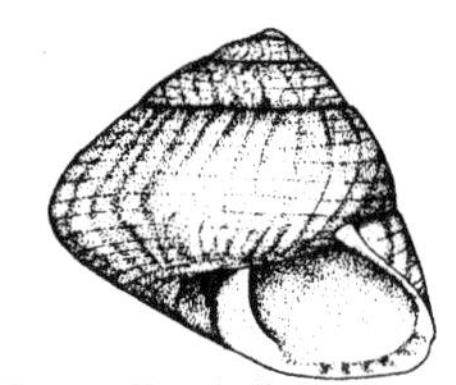
Fig. 46. Top shell.

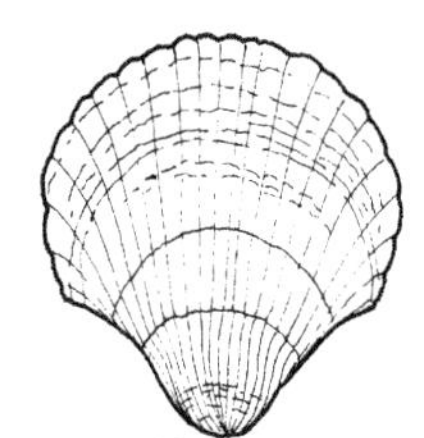
Fig. 47. Cockle.

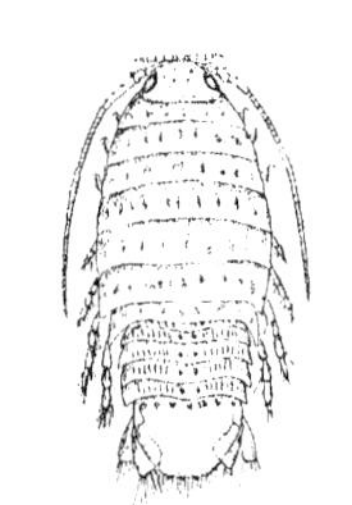
Fig. 48. Marine isopod.

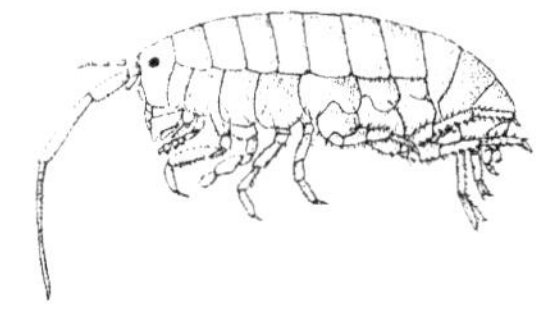
Fig. 49. Gammarid amphipod.

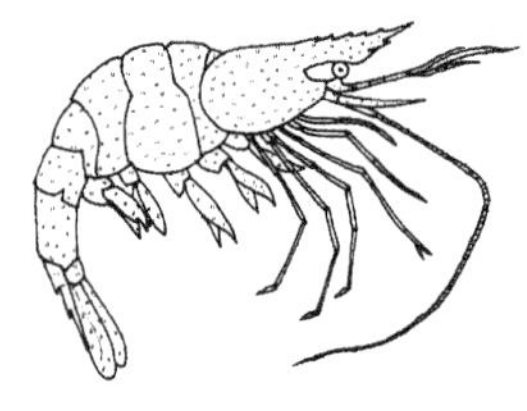
Fig. 50. Prawn.

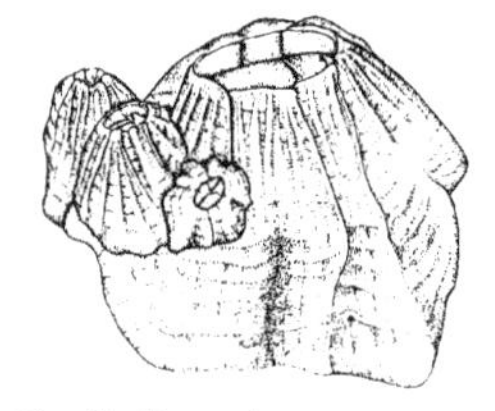
Fig. 51. Barnacle.

whelks, and top shells (fig. 46), and are dealt with by Graham (1988). Sea slugs, which lack shells, may be seen below the water line in rock pools and on rocky shores (pl. 1.1). Most species are covered by Thompson (1988). The most common molluscs on sandy substrates are bivalves such as razor shells (pl. 1.2) and cockles (fig. 47) in which the shell consists of roughly symmetrical halves (valves) hinged along the edge. Tebble (1976) includes all British species.

Crustacea, especially isopods, crabs and amphipods, are common on both rocky and sandy shores. Isopods are flattened from top to bottom, have five to seven pairs of jointed legs and are often common on sandy beaches and amongst seaweeds on rocky shores (fig. 48). Naylor (1972) describes the coastal British species. Gammaridean amphipods are flattened from side to side and have seven pairs of jointed legs (fig. 49). They are widely found on all types of shore. Lincoln (1979) covers the 250 or so British species. Decapods have five pairs of legs and include crabs (pl. 1.3), prawns (fig. 50) and shrimps. Shrimps and prawns are included in Smalden (1994), while crabs are dealt with by Crothers & Crothers (1988) and Ingle (1996). Barnacles are sessile organisms covered in calcareous plates (fig. 51). They are found encrusting rock faces, especially on the upper shore near the high water mark. Southward (1976) identifies the British species.

Echinoderms are radially symmetrical animals, including sea urchins (more or less spherical animals covered in spines when alive: pl. 1.4), starfish (broad-armed, star-shaped animals: pl. 1.5) and brittlestars (slender-armed, star-shaped animals: pl. 1.6). Several species are found in shallow water on both sandy and rocky shores. Hayward (1994, NH 21) identifies common species found on sandy shores, while the field guide by Picton (1993) covers shallow-water species.

Although few insects are associated with marine habitats, some flies and midges can be found amongst seaweed and detritus, and some beetles live under strand-line debris and burrow into sand at the base of dunes. These are generally thought of as terrestrial species and may be covered in texts dealing with such groups (see the section on terrestrial insects on pp. 39–42).

Freshwater invertebrates

The types of freshwater invertebrates found differ between standing and running waters, and between those active at the surface, in the water column and on the bed of the water body. The Linnean Society of London and the Freshwater Biological Association (addresses p. 98) produce keys to many aquatic invertebrate groups. There are several other guides available, including keys to the major groups (Croft, 1986), a field guide to common freshwater invertebrates (Fitter & Manuel, 1994), and a small guide covering the common animals found in ponds and streams (Olsen and others, 2001). In addition, Goddard (2000) has

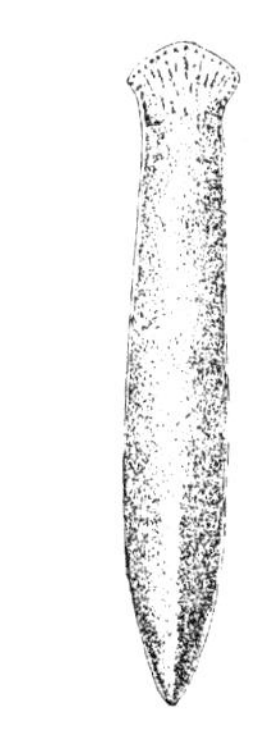
Fig. 52. Flatworm.

produced a guide to many adult insects found around freshwater, for anglers who wish to produce flies for fishing. Armitage and others (1979) list works dealing with the identification of freshwater invertebrates, although this is getting a little out of date. Some insects have larvae that are aquatic, but are terrestrial as adults and may be included in texts dealing with terrestrial insects (see pp. 39–42). Brackish water animals are also dealt with in depth by Barnes (1994) which covers all those likely to be found in Northwestern Europe. Below, we discuss some of the major groups of larger invertebrates found in freshwaters, and suggest aids for their identification. Unless otherwise stated, the identification guides mentioned contain keys to species. Some animals are more amenable than others to investigation if you are relatively inexperienced. For example, pond snails, isopods, gammarids and many insect larvae (such as stoneflies, mayflies, dragonflies and caddisflies) are relatively easy to catch, observe and identify and offer the opportunity for a wide range of investigations.

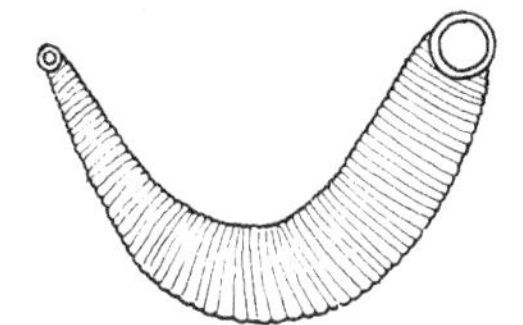
Fig. 53. Leech.

Several types of worms are found in freshwater. Common groups are the Platyhelminthes (flatworms: fig. 52) and Annelida (segmented worms). The annelids include several earthworm and potworm species as well as Tubificidae (river-worms) and leeches. River-worms are adapted to low oxygen conditions and can be found head down in the mud, waving their bodies to create currents to enhance the oxygen collecting process. Leeches have a large sucker at the hind end that enables them to hold onto surfaces, and a smaller sucker round the mouth which is used to grip prey while they suck its blood (fig. 53). Most leeches are blood feeders. There are keys for flatworms by Reynoldson & Young (2000), for leeches by Elliott & Mann (1979), and for other worms by Brinkhurst (1963).

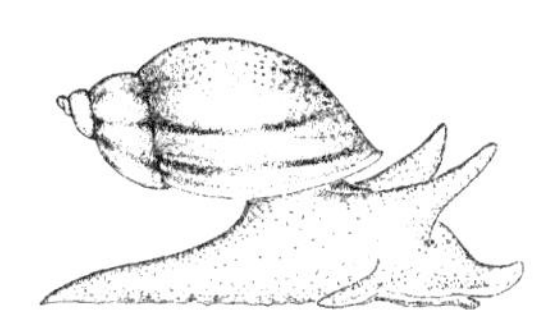
Fig. 54. Pond snail.

Molluscs found in freshwaters include both gastropods and bivalves. The former are snails (fig. 54) and limpets that are active on stones, aquatic plants and even the underside of the surface film. They graze algae and plant tissue. Bivalves are filter feeders that live on or in the substrate at the bottom of rivers, streams, lakes and ponds (for example, freshwater mussels: fig. 55). Freshwater gastropods are covered by Macan (1994), while Ellis (1978) and Killeen and others (in press) deal with the bivalves.

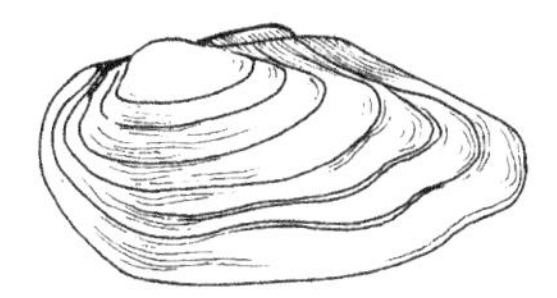
Fig. 55. Freshwater mussel.

Crustacea are common in freshwater. Many of the larger species belong to the Malacostraca and include isopods (which are similar to woodlice) and amphipods such as gammarids (freshwater shrimps). The former trundle about on the beds of rivers and ponds and on aquatic vegetation (fig. 56), while the latter are more active, scooting about near the bottom (fig. 57). Both are largely scavengers and are covered by Gledhill and others (1993). Cladocera include the water fleas (fig. 58) and are much smaller animals found in large numbers in still waters, often as members of the plankton. They comprise two groups: the Calyptomera are filter-feeders and the Gymnomera are predators. Both are dealt with by Scourfield & Harding (1966).

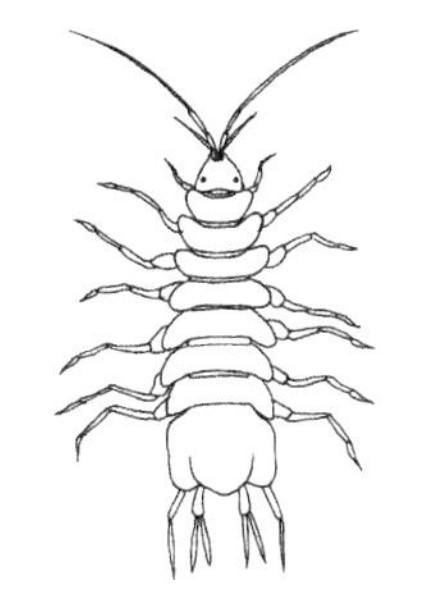
Fig. 56. Freshwater isopod.

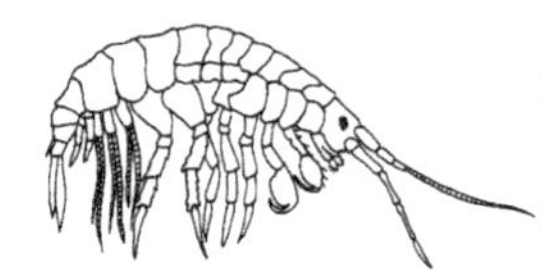

Fig. 57. Gammarus.

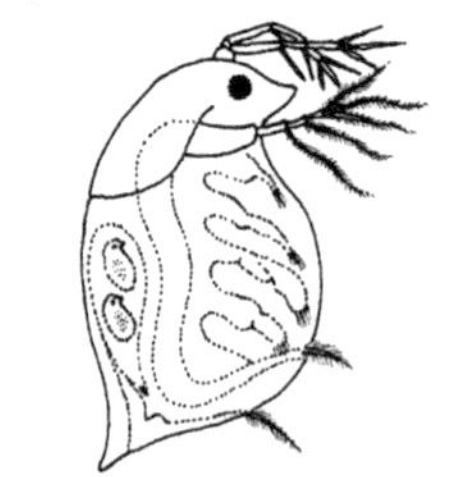

Fig. 58. Water flea (a cladoceran).

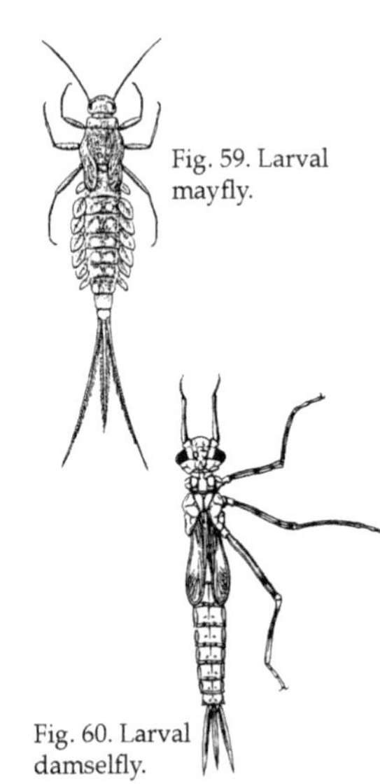

Fig. 59. Larval mayfly.

Fig. 60. Larval damselfly.

Occasionally spiders are found in and on the water, including the raft spider (pl. 2.1) and water spider. These are covered in the major work by Roberts (1993) and his field guide (Roberts, 1995). Water mites are often abundant on water plants, on the bottom or actively swimming (pl. 2.2). They are more challenging to identify, but there is an old but well-illustrated monograph on water mites by Soar & Williamson (in three volumes, 1925, 1927 and 1929) which may be used in conjunction with a more up-to-date checklist by Gledhill & Viets (1976).

Insects are frequently found on the substrate at the bottom of the water column, free swimming within the water, on the surface film, amongst aquatic vegetation or flying freely overhead. Some insects have aquatic larvae and terrestrial adults and may be covered by different texts for different life stages. The Royal Entomological Society (address p. 99) produces Handbooks for the Identification of British Insects that cover some aquatic groups: parts in volume 1 deal with caddis, volume 2 with true bugs, volumes 4 and 5 with beetles, and volumes 9 and 10 with true flies.

Of those insects with aquatic larvae and terrestrial adults, mayflies (Ephemeroptera: pl. 2.3 and fig. 59) are covered by Harker (1989, NH 13); dragonflies and damselflies (Odonata: pl. 2.4 and fig. 60; see also fig. 82) in a key by Miller (1995, NH 7), the field guide by Brooks (1997), and a wall chart by Askew (1988); stoneflies (Plecoptera: fig. 61) by Hynes (1977); caddis (Trichoptera: pl. 2.5 and fig. 62) by Wallace and others (1990) and Edington & Hildrew (1995); and alderflies (Megaloptera: fig. 63) and some lacewings (Neuroptera) by Elliot (1996). Larvae of dragonflies, alderflies, aquatic lacewing larvae and some stoneflies are carnivorous, whilst mayfly larvae and other stonefly larvae are mainly herbivorous, and caddis larvae cover most feeding types with some species being predatory while others are decomposers. Stonefly larvae are usually found in running water and are intolerant of organic pollution, as to a lesser extent are mayfly larvae. This has led them to be used as biological indicators of pollution (Hellawell, 1986) and makes them useful subjects to study. Some caddis larvae (cased caddis) cover themselves with a protective case made from sand grains or plant material.

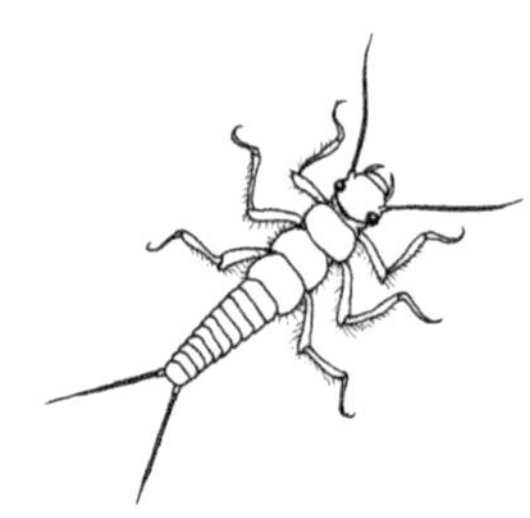

Fig. 61. Larval stonefly.

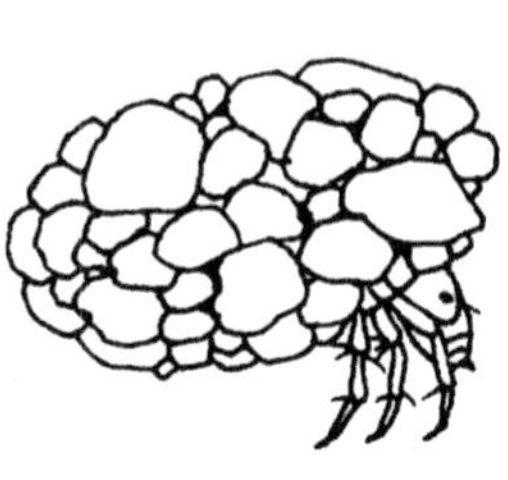

Fig. 62. Larval cased caddis.

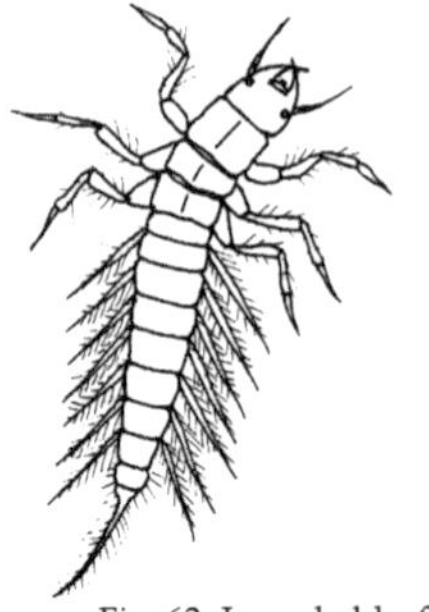

Fig. 63. Larval alderfly.

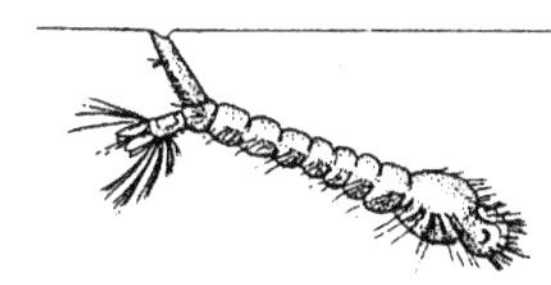
Fig. 64. Mosquito larva.

Fig. 65. Midge larva.

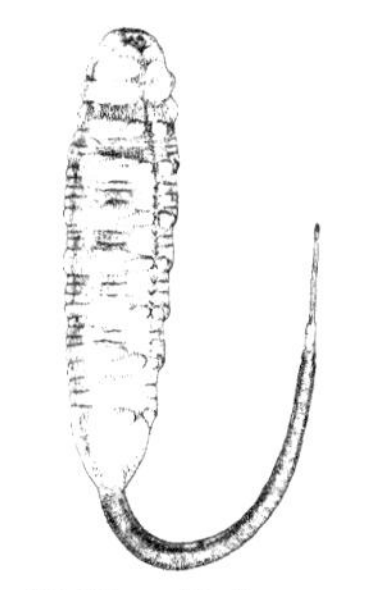
Fig. 66. Hoverfly larva.

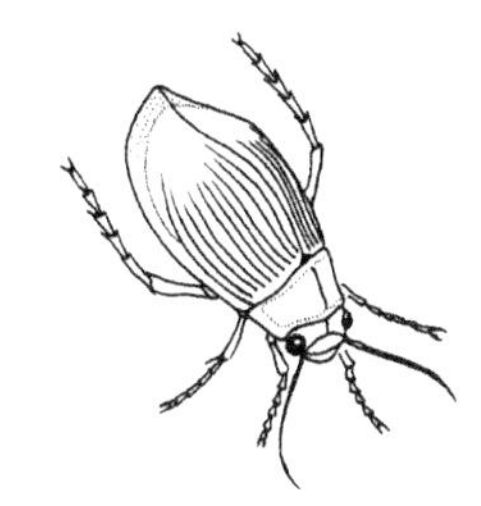
Fig. 67. Diving beetle.

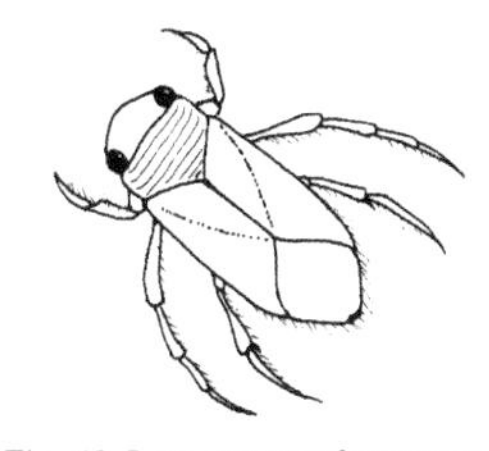
Fig. 68. Lesser water boatman.

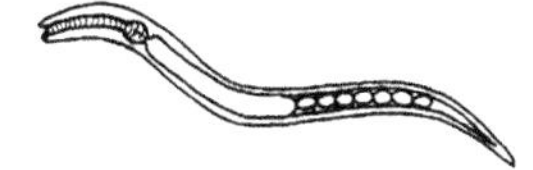
Fig. 69. Eelworm.

True flies (Diptera) are also found only as larvae within freshwater. Many fly larvae feed on small particles they filter out from the water, either at the surface film or deeper in the water. This order is covered by a range of keys including Smith (1989) on groups of fly larvae (and eggs and pupae), Snow (1990, NH 14) on mosquitoes (pl. 2.6 and fig. 64), Cranston (1982) on non-biting midges (Chironomidae), and Disney (1999) on meniscus midges (Dixidae: fig. 65) and trickle midges (Thaumaleidae). Some hoverflies also have aquatic larvae (fig. 66). Gilbert (1993 NH 5) and Stubbs & Falk (1996) give further information.

Other insect orders that may be found as both adults and larvae in freshwaters include true bugs (Hemiptera) and beetles (Coleoptera). Many water beetles, such as whirligig (Gyrinidae: pl. 2.7) and diving beetles (Dytiscidae: fig. 67), are carnivorous, as are some water bugs, such as backswimmers (Notonectidae: pl. 2.8) and pondskaters (Gerridae: pl. 2.9). Lesser water boatmen (Corixidae: fig. 68) on the other hand, are herbivorous. Keys to the families of both true bugs and beetles have been written by Unwin (2001, and 1988, respectively). The only colour-illustrated field guide taking bugs to species is long out of print (Southwood & Leston, 1959), but is available on CD-ROM from Pisces Conservation Ltd (address p. 98). However, Savage has produced a key to adult bugs in freshwaters (1989) as well as a key covering lesser water boatman larvae (1999). The beetles are covered by both a field guide (Harde, 2000) and a key to adults by Friday (1988).

Terrestrial invertebrates

Terrestrial habitats contain a wide variety of invertebrate animals, especially insects. This large number of potential species means that identification can be difficult. The Linnean Society of London and The Royal Entomological Society (addresses pp. 98, 99) have published keys taking some groups to species, and Tilling (1987) has produced a key to the major groups. Below, we consider some of the major groups of larger invertebrates found in terrestrial habitats, and suggest aids for their identification. Unless otherwise stated, the identification guides mentioned contain keys to species. Some animals are more amenable than others to investigation if you are relatively inexperienced. For example, snails, woodlice, some spiders, harvestmen, millipedes, centipedes, grasshoppers, butterflies, moths, ants, and some flies and beetles (such as hoverflies, ground beetles and ladybirds) are relatively easy to catch, observe and identify and offer the opportunity for a wide range of investigations.

Several types of worms are found in terrestrial sediments, including eelworms (Nematoda) and members of the Annelida (segmented worms). Eelworms are very numerous and often very small (fig. 69). Many species are as yet undescribed, and those that have been described tend to

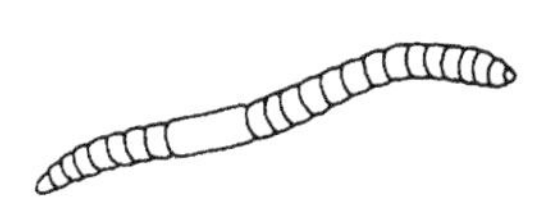

Fig. 70. Earthworm.

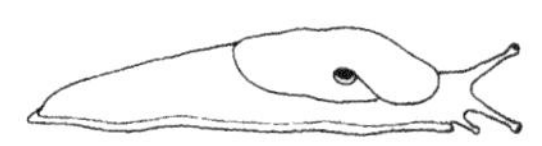

Fig. 71. Slug.

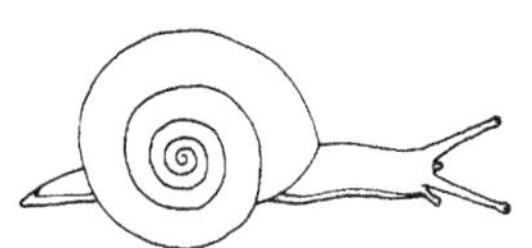

Fig. 72. Snail.

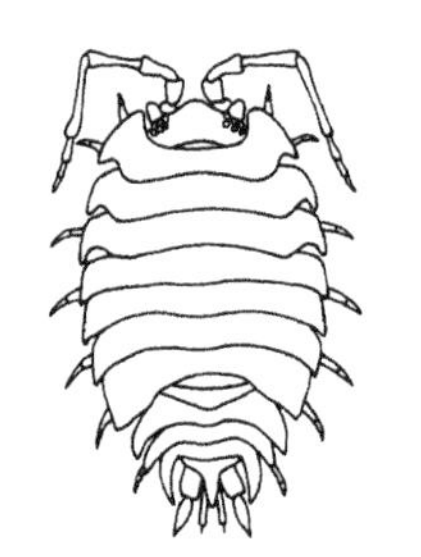

Fig. 73. Woodlouse.

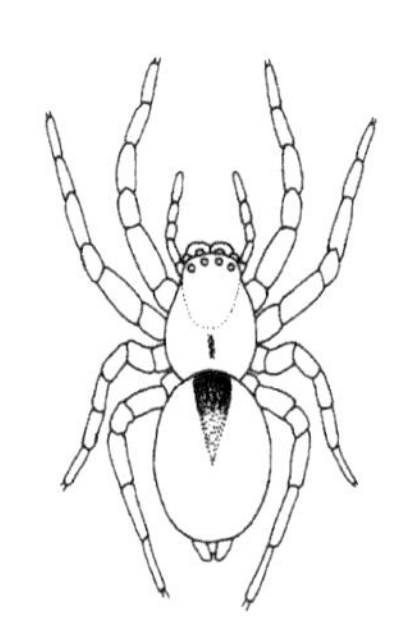

Fig. 74. Spider.

be of economic interest as pests of crops and parasites of humans and our domesticated animals. There is no comprehensive up-to-date taxonomic work on eelworms, and their identification is a task for an expert. Annelid worms, which include earthworms (fig. 70) and potworms, are cylindrical soft-bodied animals with obvious segmentation. Most earthworms inhabit the soil, feeding on decaying plant material, and are important in the breakdown of organic material and in soil mixing. Sims & Gerard (1985) cover British earthworms, while Edwards & Lofty (1977) give a key to the common genera of earthworms.

Slugs and snails are common terrestrial molluscs. Slugs (fig. 71) are distinguished from snails (fig. 72) by having either no shell, or such a small shell that it would be impossible for the animal to fully retreat into it. Most species are herbivorous, nocturnal or active only when wet, and all are restricted by the danger of desiccation. Snails benefit by having a waterproof shell, but slugs are able to burrow deeper into the soil. There are identification guides to slugs (Cameron and others, 1983) and land snails (Cameron & Redfern, 1976), as well as a field guide covering both slugs and land snails (Kerney and Cameron, 1979).

The major terrestrial representatives of the Crustacea are the woodlice (fig. 73). These animals are found in aggregations under logs and in other decomposing plant material such as compost heaps. Woodlice may be important in recycling dead plant material since they feed mainly on dead and decaying vegetable matter, which they break down and excrete as small fragments that then decompose readily. Identification is covered by Hopkin (1991) and Oliver & Meechan (1993).

Terrestrial Chelicerata include spiders, mites, ticks, harvestmen and pseudoscorpions. Most species are predatory, although some mites feed on plant material. Spiders (fig. 74) and harvestmen (fig. 75) are large predators that can be found within vegetation or on the surface of the ground, either lying in wait for prey or actively hunting it down. For spiders, a key to families by Jones-Walters (1989) and a major key to species by Roberts (1993) are supplemented by a field guide (Roberts, 1995). There is a key to harvestmen by Hillyard & Sankey (1989). Pseudoscorpions are small, aggressive, predatory, scorpion-like animals (although they lack the sting), and may be found in leaf litter and grassland (fig. 76). They are covered in a key by Legg & Jones (1988). Mites and ticks may be more difficult to

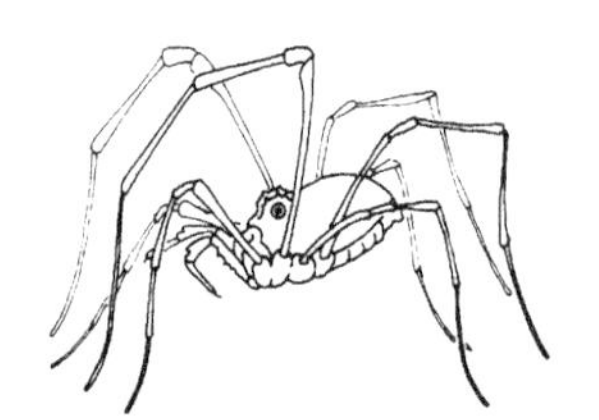

Fig. 75. Harvestman.

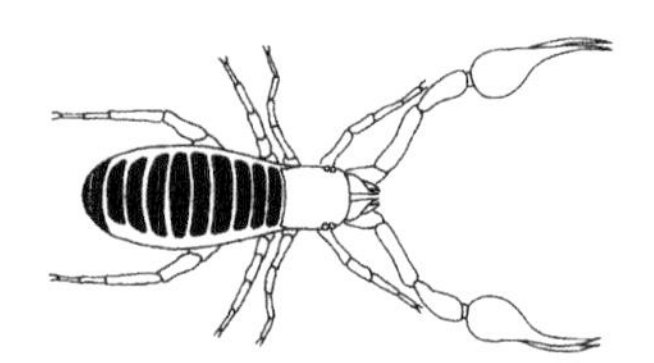

Fig. 76. Pseudoscorpion.

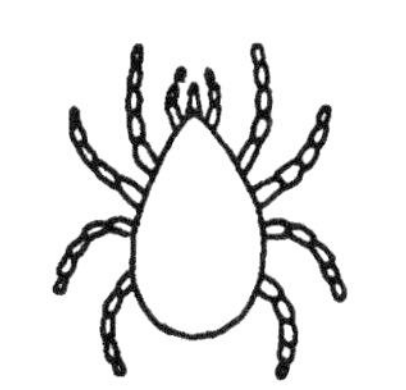
Fig. 77. Mite.

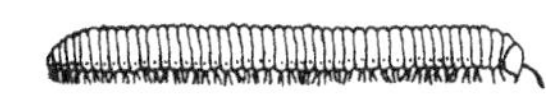
Fig. 78. Snake millipede.

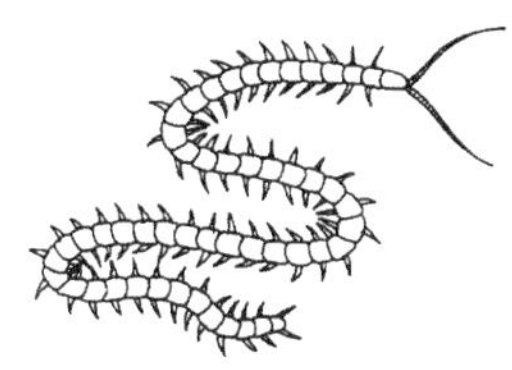
Fig. 79. Geophilomorph centipede.

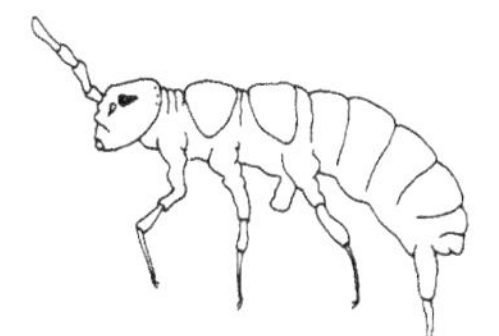
Fig. 80. Springtail.

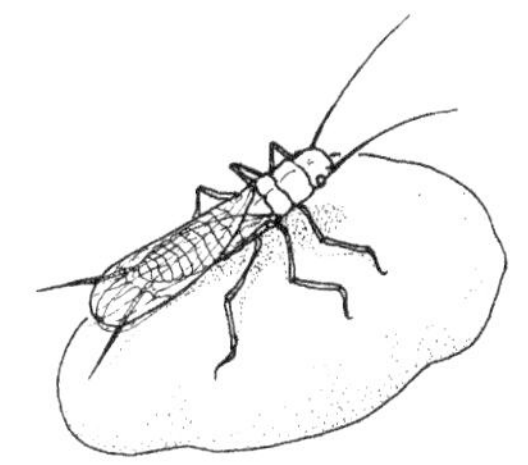
Fig. 81. Stonefly.

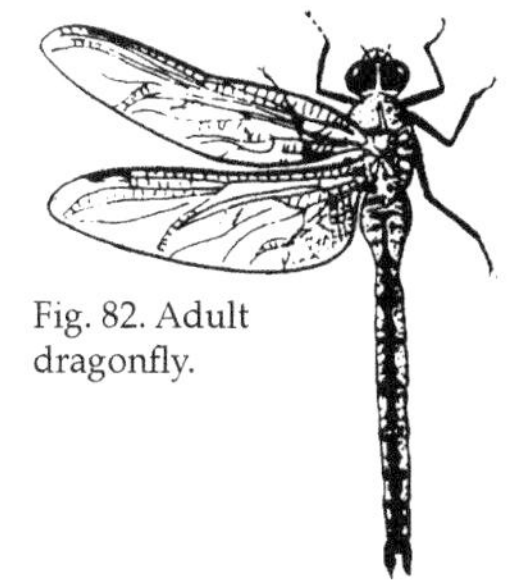
Fig. 82. Adult dragonfly.

identify. Ticks are parasitic blood-sucking animals and are keyed by Hillyard (1996). Mites are much more diverse in habits and there is no recent comprehensive treatment (fig. 77). McDaniel (1979) gives keys to families and illustrations of selected species.

The Myriapoda comprise two classes: the Diplopoda (millipedes) and the Chilopoda (centipedes). Millipedes are decomposers, contributing to the breakdown of vegetable matter, while centipedes are predators. Both are found on the soil surface, under logs and stones and in the soil. There are several common groups of millipedes including snake millipedes (fig. 78), pill millipedes (pl. 3.1) and flat backed millipedes (pl. 3.2). All are covered by Blower (1985). Eason (1964) deals with centipedes. More recent works identify different groups of centipedes: the Geophilomorpha (fig. 79), which are long, thin and slower moving, are dealt with by Keay (1995) and Barber & Keay (1995), while the Lithobiomorpha (pl. 3.3), which are stouter and fast moving, are covered by Barber (1996).

A wide range of insects inhabit terrestrial habitats, although some have aquatic larvae and terrestrial adults (discussed in the freshwater section, pp. 34–37). Corke (1999) gives details of the keys available for many insect groups, including the Handbooks for the Identification of British Insects published by The Royal Entomological Society (address p. 99) that cover some groups: see parts in volume 1 for small orders such as grasshoppers, volume 2 for true bugs, volumes 4 and 5 for beetles, volumes 6, 7 and 8 for ants, bees, wasps and allied insects, and volumes 9 and 10 for true flies. There are several field guides, for example by Chinery (1986), Chinery (1993), and Zahradník (1998*a*), and many of the Naturalists' Handbooks cover the insects that occur in particular habitats, such as on nettles (Davis, 1991, NH 1), thistles (Redfern, 1995, NH 4), cabbages (Kirk, 1992, NH 18), docks (Salt & Whittaker, 1998, NH 26) and cherry trees (Leather & Bland, 1999, NH 27), and under logs and stones (Wheater & Read, 1996, NH 22).

Springtails (Collembola) are common small, flightless, soil insects that feed mainly on decomposing plant material and fungal hyphae (fig. 80). Some species have a 'spring' at the hind end that is released when the animal is disturbed, catapulting it into the air as an escape response. Hopkin (in press) keys out all British species.

Both mayflies and stoneflies have aquatic larvae, but flighted adults. Mayflies (pl. 2.3) belong to the order Ephemeroptera, named for the short period during which the adults live (up to a few days or even only a few hours). Mayflies are unique amongst flying insects in that they moult once winged. They are weak fliers and need calm conditions for swarming. Harker (1989, NH 13) covers the British species. Stoneflies (fig. 81) are also weak fliers and short-lived as adults (surviving for two to three weeks). They may be found crawling on stones at the edges of streams. Hynes (1977) covers British stoneflies.

Dragonflies (fig. 82) and damselflies (Odonata) also

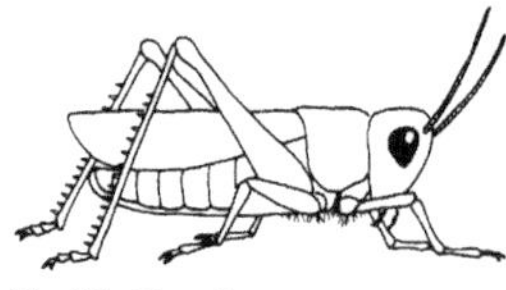
Fig. 83. Grasshopper.

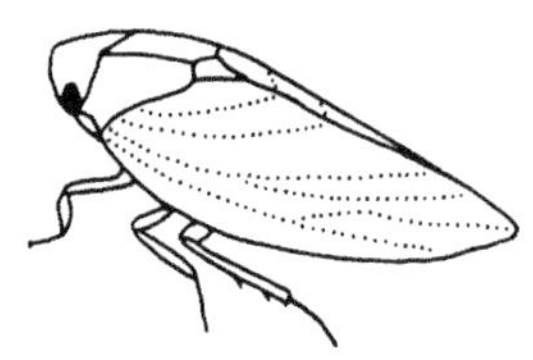
Fig. 84. Leafhopper.

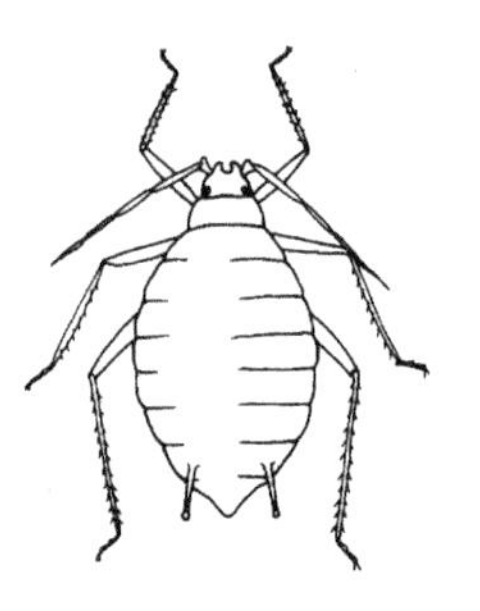
Fig. 85. Aphid.

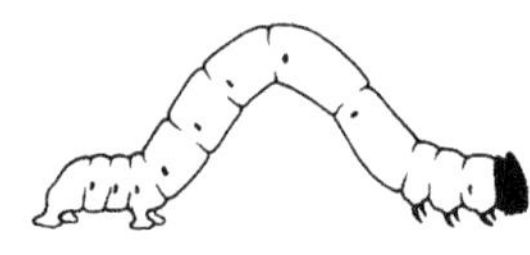
Fig. 86. Caterpillar.

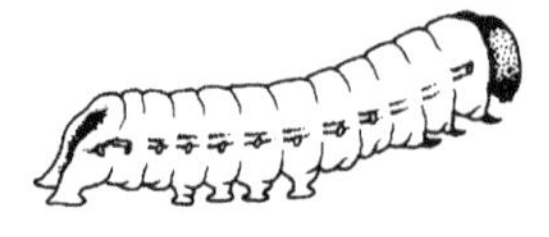
Fig. 87. Caterpillar.

have aquatic larvae and flighted adults. They are strong fliers (especially the dragonflies) and may be seen darting and hovering near open water or resting on emergent vegetation. They are opportunistic predators that often take prey on the wing. Texts for their identification include the key by Miller (1995, NH 7), the field guide by Brooks (1997) and the wall chart by Askew (1988).

The Orthoptera include grasshoppers (fig. 83) and bush-crickets (pl. 3.4) and are nearly all herbivores, using their tough mouthparts to chew leaves and other plant material. Identification of all British grasshoppers and bush-crickets is covered by Brown (1990, NH 2), while these and related species of Orthoptera are dealt with by Marshall & Haes (1990). Many species of grasshoppers and crickets produce sounds by stridulation: rubbing together a series of teeth (the file) and a thickened ridge (the scraper). This pair of structures may be located on the wings, or the file on the hind leg and the scraper on the wings. More details, and a sound guide, have been produced by Ragge & Reynolds (1998) and Burton & Ragge (1988) respectively.

The true bugs (Hemiptera) are a diverse group of animals that have piercing and sucking mouthparts. Many species feed on plant material, although some of the sub-order Heteroptera suck the blood of vertebrates or invertebrates. The Heteroptera (such as shield bugs, pl. 3.5), are distinguished by having forewings that are partially hardened, producing a partial leathery cover for the membranous hind wings. The two other sub-orders, the Auchenorrhyncha including planthopppers and leafhoppers (fig. 84) and the Sternorrhyncha including greenflies or aphids (fig. 85), have wings that are uniform in texture. In winged aphids the forewings are entirely membranous, whereas in some planthoppers they are entirely leathery. Unwin (2001) has produced a key to the families of true bugs, and there is an old but excellent illustrated field guide to the species of Heteroptera by Southwood & Leston (1959). Other works concentrate on individual groups, such as aphids (Blackman, 1974).

Lacewings (Neuroptera) have carnivorous larvae and adults (pl. 3.6), although the latter may rarely feed. Adults have membranous wings that are held tent-like over the body when at rest. Plant (1997) covers the identification of adult lacewings and allied insects, while Elliott (1996) deals with the adults of those with aquatic larvae and gives a key to all families.

Some of our most colourful insects are the day-flying Lepidoptera. These are mainly butterflies (pl. 3.7) and skippers, although some moths are also diurnal. Some nocturnal moths are also vividly patterned (pl. 3.8).The caterpillars of butterflies and moths (figs. 86 and 87) are mainly herbivorous chewers of plant material, with a few exceptions, some of which are household pests feeding on wool and other animal material. Most adults are liquid feeders, using the proboscis to take nectar and other fluids, although in some species the mouthparts are not complete

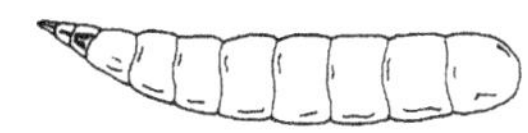

Fig. 88. Fly larva.

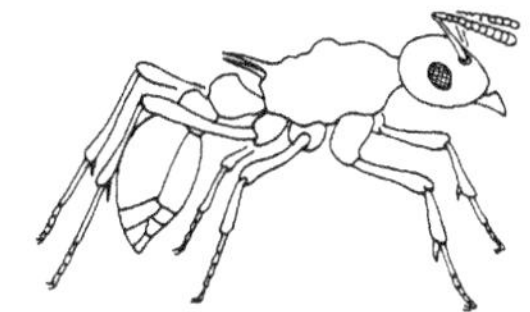

Fig. 89. Ant.

Fig. 90. Solitary wasp.

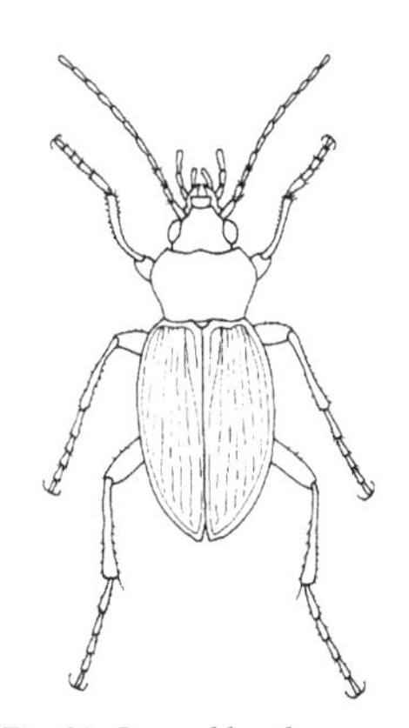

Fig. 91. Ground beetle.

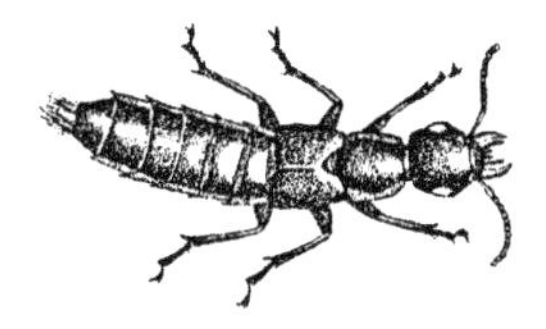

Fig. 92. Rove beetle.

and the adults do not feed. Species in one superfamily (the Micropterigidae) possess functional jaws as adults and feed on pollen. The wings of Lepidoptera are covered by scales which often contain the colour pigments. In addition to a major work on the Lepidoptera (with only some volumes as yet published) edited by Emmet and others (1983 onwards), there are many other identification guides, including field guides to adult butterflies by Tolman & Lewington (1997) and to caterpillars by Carter & Hargreaves (1986), and colour guides to the larger moths by Skinner (1998) and to larvae by Porter (1997). Smaller species of moths are covered by other texts such as Goater (1986), who deals with pyralid moths.

The true flies (Diptera) are a very diverse group of insects, including species with larvae that are predators (such as hoverflies), herbivores (such as mosquitoes), decomposers (such as dung-flies), or parasitic on plants (such as gall-forming midges). The adults have one pair of functional wings, with the second pair (the hind wings) reduced to halteres (small balancing devices that assist in flight). Unwin (1981) has produced a key to families and Colyer & Hammond (1968) is an old text containing keys to families, coloured illustrations and notes on biology. Particular groups of flies are covered in more detail by texts such as Gilbert (1993, NH 5) and Stubbs & Falk (1996) for hoverflies (pl. 4.1), Snow (1990, NH 14) for mosquitoes (pl. 2.6), Redfern & Askew (1992, NH 17) for gall-forming flies, and Erzinçlioğlu (1996, NH 23) for blowflies (pl. 4.2 and fig. 88).

The Hymenoptera include the nectar- and pollen-feeding bees (pl. 4.3), the omnivorous ants (fig. 89) and the carnivorous social and solitary wasps (pl. 4.4 and fig. 90 respectively), as well as a range of parasitic groups such as ichneumon flies and gall-forming wasps. Winged adult Hymenoptera have two pairs of membranous wings linked together with the larger front pair attached by a row of small hooks on the front edge of the hind pair. Keys for particular groups of Hymenoptera include Willmer (1985) for the genera of bees, ants and wasps; Wright (1990) for genera of sawflies; Yeo & Corbet (1995, NH 3) for solitary wasps; Prŷs-Jones & Corbet (1991, NH 6) for bumblebees; Redfern & Askew (1992, NH 17) for gall-forming wasps, and Skinner & Allen (1996, NH 24) for ants. A small field guide by Zahradník (1998*b*) covers common bees and wasps.

There are more than 4000 species of beetles in Britain. Beetles are highly variable in both size and shape, but in adult beetles the hardened fore wings cover (or partially cover) the abdomen and protect the membranous hind wings. Beetles include predators such as ground beetles (fig. 91), plant feeders such as weevils (pl. 4.5), and decomposers such as dung beetles. Other common families include rove beetles (fig. 92), leaf beetles (pl. 4.6) and ladybirds (pl. 4.7). The standard key to beetle families and species is the work by Joy (1932), which is now getting a little old and has been supplemented by Hodge & Jones (1995). There are keys to families by Unwin (1988) and a field guide to common species by Harde (2000). Particular groups are covered by

Forsythe (2000, NH 8) for ground beetles, Majerus & Kearns (1989, NH 10) for ladybirds, and Morris (1991, NH 16) for weevils.

4 Describing data

An investigation often yields a series of observations, such as counts of insects on flower heads, timings of bee visits to flowers, sizes of a mollusc species in different zones of a rocky shore, ranked scores of degree of decay of logs, or records of whether an invertebrate is a juvenile or adult. Descriptive statistical methods help us to summarise the data and identify patterns so that we can more easily understand what is happening and more efficiently communicate the results to other people. There are a variety of methods for displaying graphical and tabular data, as well as summary statistics that reduce a list of numbers to one or two values that define properties of the data set (such as the average value, or the most frequently occurring value). For example, we might wish to identify one or more of the following aspects:

- how often particular data values are obtained;
- the average value of a series of observations;
- the extent to which a series of observations vary;
- the population size or density of a given group of organisms in a particular area;
- whether the distribution of organisms in space is even, random or clumped;
- how diverse communities are in terms of species composition;
- how similar communities are in terms of species composition.

Table 2. *Frequency table for number of eggs laid by hoverflies on brassica crops.*

Size class of number of eggs per cabbage	Frequency
10–14	/
15–19	/
20–24	//
25–29	////
30–34	//
35–39	~~////~~ //
40–44	~~////~~
45–49	~~////~~
50–54	//
55–59	//
60–64	//
65–69	/
70–74	
75–79	//
80–84	
85–89	//
90–94	
95–99	
100–104	/
105–109	
110–114	
115–119	
120–124	
125–129	
130–134	
135–139	
140–144	/

These situations are dealt with in more detail within this chapter. They are illustrated by reference to a range of examples and are supported by boxed worked examples. The data used are based on real examples (although sometimes the data have been manipulated to simplify the calculations and so the results are not necessarily indicative of the true situation). The calculations are often based on numbers given to several decimal places. This is to avoid errors due to numbers being rounded up or down to a small number of decimal places. However, large numbers of decimal places are usually avoided when displaying the final results. The boxed worked examples demonstrate the number of decimal places that should be recorded in your final results. There are few commercial statistical packages that perform all these tests. An easy to use Excel workbook is available from the authors (address p. 100) that illustrates the calculations using the data from the worked examples in this book and can be used to analyse your own data. See Appendix I when calculating statistics by hand.

Examining frequencies

If we count the numbers of eggs laid by 40 hoverflies on brassica crops, we obtain 40 counts, or data points.

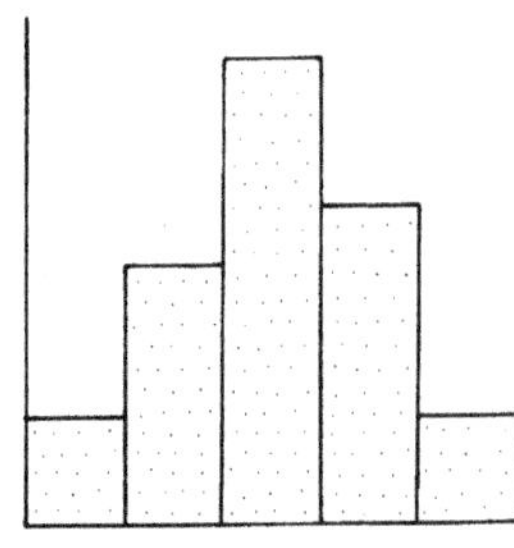
Fig. 93. Frequency histogram.

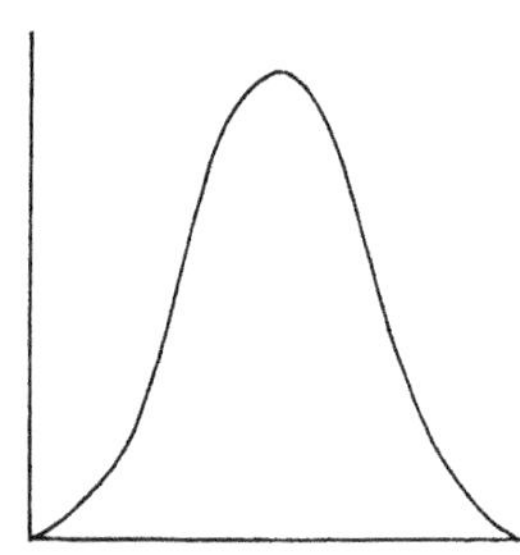
Fig. 94. Normal distribution.

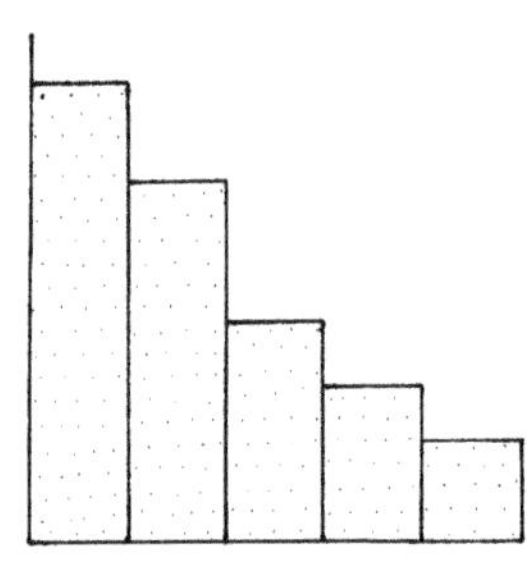
Fig. 95. Positively skewed distribution.

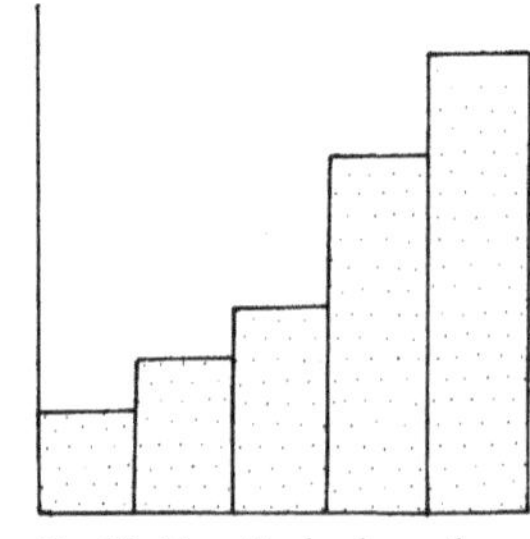
Fig. 96. Negatively skewed distribution.

To help us appreciate the range (from the smallest to the largest value) and the distribution of values (how many low and high values there are compared with how many there are in the middle of the range), it is useful to examine a frequency distribution, either in tabular or graphical form. This simply involves listing in a frequency table (table 2), or plotting in a frequency histogram (fig. 93), all possible values of the counts or measurements, in rank order, with the frequency at which each occurs (box 4.1). The shape of such a frequency histogram is called the distribution. When we plot data such as these, especially if data points are numerous, they commonly form a symmetrical, bell-shaped curve (fig. 94). The bulk of the data points lie around the middle, and there are relatively few extremely low and extremely high values. Distributions like these are known as normal distributions.

Data do not always follow a normal distribution. Many other shapes are possible. Some may be positively skewed so that there are relatively large numbers of low values and fewer larger values (fig. 95). This is particularly common when counting relatively rare items, for example the numbers of hoverflies visiting flowers in 10-minute intervals. Others are negatively skewed, where there are more larger values and fewer smaller ones (fig. 96). If we measured the height of trees in a plantation that was planted on bare ground fifteen years ago, most trees would have grown to similar heights. Although there would not be many very tall trees, since no trees at all were present fifteen years ago, there could be some smaller saplings growing underneath the planted trees. It is important to be aware of the shape of the frequency distribution of your data, since some statistical techniques require data to approximate to a normal distribution (Chapter 5).

A worked example of the production of a frequency table and histogram for data on egg laying in hoverflies is shown in box 4.1. Counts of invertebrates, such as the numbers of grasshoppers found in several samples from a meadow (Brown, 1990, NH 2), can also be displayed as frequency tables or histograms, as can measurements such as heights of plants or time spent foraging.

Finding the average value

One of the most straightforward descriptive statistics is the average value. The most commonly used average is the arithmetic mean, which is simply the sum of all of the values, divided by the total number of data points in the sample. The mean is a measure of central tendency, and is appropriate when the data are approximately normally distributed measurements. Note that the mean is expressed to one more decimal place than the data that were used to calculate it. Thus if we have several counts of insects on nettles of 45, 34, 52 and 27, then the mean number per nettle plant is 39.5. Similarly if we have measured the lengths of logs on a

Box 4.1 Frequency tables and histograms

An investigation of egg laying in hoverflies involved counting eggs on brassica crops. In order to examine and visualise these data, frequency tables and histograms may be used.

1. Record the data (number of eggs per cabbage).

31	86	20	41	79	75	88	58	56	37	60	41	45	47	18	32	43	50	13	47
40	35	38	38	104	60	144	40	69	39	54	35	48	26	29	29	46	21	35	28

2. Organise the data in order and place into a frequency table (in this case using classes of 5 in order to avoid too many empty cells).

See also table 2.

Size class	10–14	15–19	20–24	25–29	30–34	35–39	40–44	45–49	50–54
Frequency	1	1	2	4	2	7	5	5	2
Size class	**55–59**	**60–64**	**65–69**	**70–74**	**75–79**	**80–84**	**85–89**	**90–94**	**95–99**
Frequency	2	2	1	0	2	0	2	0	0
Size class	**100–104**	**105–109**	**110–114**	**115–119**	**120–124**	**125–129**	**130–134**	**135–139**	**140–144**
Frequency	1	0	0	0	0	0	0	0	1

3. Interpret the results.

It may be useful to use the frequency table to draw a frequency histogram.

The data show a slightly skewed distribution with one or two counts that are quite high. However, the majority of the counts lie around the middle of the distribution. We can see that hoverflies most commonly laid between 35 and 39 eggs. Gilbert (1993, NH 5) gives further information about egg laying in hoverflies, Rotheray (1989, NH 11) gives details about hoverflies that feed on aphids, and Kirk (1992, NH 18) discusses insects found on brassica crops.

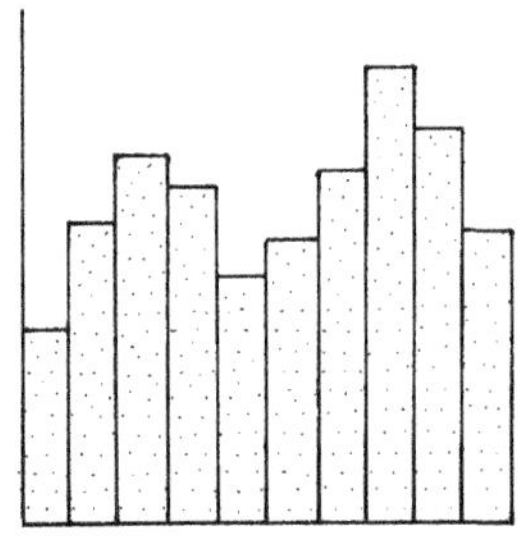
Fig. 97. Bimodal distribution.

woodland floor as 0.45 m, 0.23 m, 0.29 m, 0.60 m and 0.19 m, then the mean length of the logs is 0.352 m.

Where we have ranked values, or data that are not normally distributed, an alternative statistic, the median, may be more representative. The median is found by placing all of the values in order (from lowest to highest) and selecting the middle value. Where there is an even number of data points, the median lies between the middle two values. An alternative statistic that is less frequently used is the mode, the most frequently-occurring value, which may be read directly from a frequency table or histogram (for example in box 4.1 the mode is 35–39 eggs per cabbage). Sometimes a distribution has more than one peak (fig. 97). If this happens, it is possible that the data belong to more than one population. For example, if in a survey of the heights of dock plants you find that the distribution is bimodal (has two peaks), you might conclude that you were actually measuring the heights of two different species, not the one you expected. It is important to examine the data before statistical testing to ensure that you are aware of any anomalies like this.

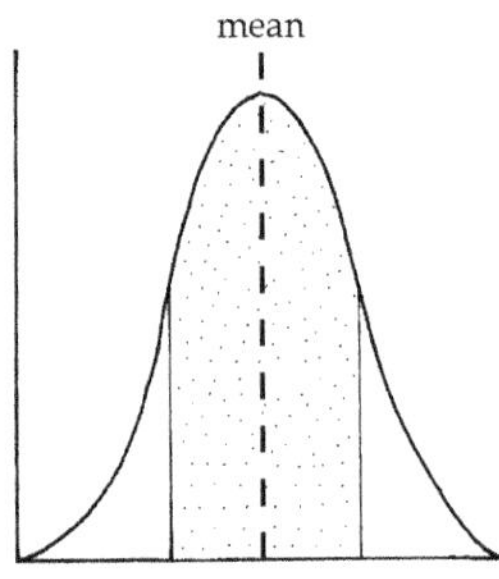

Fig. 98. Normal distribution showing mean and standard deviations.

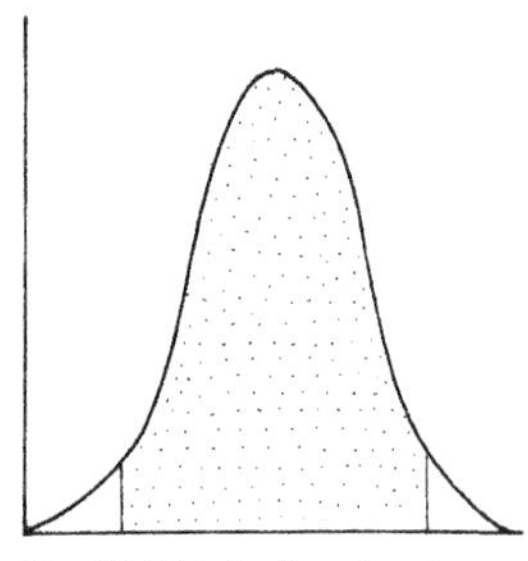
Fig. 99. Distribution showing 95% of data.

Examining variation

As well as a measure of central tendency, such as the mean or median, we need some indication of the variation of values within a data set. This can be displayed in a frequency table or histogram, or it can be represented by a summary statistic. A measure of variation appropriate for normally distributed data is the standard deviation, which is given the symbol s. When considering a frequency distribution, the standard deviation is the distance along the x axis (horizontal axis) either side of the mean that contains approximately 68% of the data points (fig. 98). A larger range of the mean plus and minus almost two standard deviations contains 95% of the data in a sample (fig. 99). Variation is sometimes expressed in terms of the variance, which is simply the square of the standard deviation (s^2). Standard deviations are recorded at one decimal place more than that of the mean. So for our earlier example of counts of insects on nettle plants, the mean was 39.5 (p. 44) and the standard deviation is 11.15. Similarly the mean lengths of logs earlier was 0.352 m (see above) and the standard deviation is 0.1704.

The standard deviation represents variation about the mean for a set of data that are normally distributed. For data that are not normally distributed, we use an alternative measure of variation about the median – the interquartile range. Just as the median is the half way point through the data ranked in ascending order, the quartiles are the two quarter way points (called Q1 and Q3) either side of the median. The range from the first quartile (Q1) at 25%, to the third quartile (Q3) at 75%, contains 50% of the data points in a sample. Often counts of animals (or plants) may be heavily skewed: for example, there may be a relatively large number of low counts and relatively few high counts (fig. 95). If this is the case, means and standard deviations are not appropriate unless the data are transformed mathematically (see Wheater & Cook, 2000, for further details). If the frequency distribution of your data is not an approximately symmetrical, bell-shaped curve (that is, unlike the distribution in fig. 94), then it may be better to use medians and the interquartile range rather than means and standard deviations.

A set of measurements can be summarised in terms of a measure of central tendency (such as the mean or median), a measure of variation (such as the standard deviation or interquartile range), and the number of measurements in the sample (n). Means and standard deviations can be displayed graphically using point charts or bar charts (p. 94), and medians and interquartile ranges can be displayed using box and whisker plots (p. 95). Chapter 6 gives more details about displaying such data.

Box 4.2 gives the formulae for the calculation of the mean and standard deviation of the distance at which territory-holding dragonflies chase off intruders. The calculation of the mean and standard deviation is useful in many circumstances where we have several measurements or counts. For example, for a given formulation of insect repellent, it would be possible to calculate the mean time over which people are protected from mosquito attack. The standard deviation could also be calculated. Snow (1990, NH 14) gives more details about mosquitoes and chemical repellents. Alternatively, one might describe the level of insect infestation on particular plants in terms of the mean numbers of herbivores (such as greenflies) per leaf, or the mean numbers of thrips per flower, together with the associated measures of variation. Rotheray (1989, NH 11) and Kirk (1996, NH 25) give details of the ecology of greenflies and thrips respectively. Some weevils lay eggs in developing flower buds (Morris, 1991, NH 16). Following a survey of several plants, the level of infestation by a weevil species could be expressed using the mean number of eggs laid per bud (plus or minus the standard deviation).

Box 4.3 uses data on estimated percentage leaf damage on dock plants as an example of the calculation of the median and interquartile range. Another situation in which the median and interquartile range of a sample may be useful is in counts of galls on leaves. There may be many leaves without any galls, and relatively few leaves with many galls. Since this results in a skewed distribution (p. 44), the median and interquartile range are appropriate measures of the central tendency and variation of the data. Redfern & Askew (1992, NH 17) give more details of galls on leaves.

Box 4.2 Mean, standard deviation, standard error and 95% confidence limits

$$[1]\ \bar{x} = \frac{\sum x}{n} \qquad [2]\ s = \sqrt{\frac{\sum x^2 - \frac{(\sum x)^2}{n}}{n-1}} \qquad [3]\ SE = \frac{s}{\sqrt{n}}$$

Where: x is each data point;
$\bar{x}$ is the mean of x;
n is the sample size of the sample;
s is the standard deviation of the sample;
SE is the standard error;
$t_{0.05[n-1]}$ instructs you to look up t at a probability of 0.05 with n–1 degrees of freedom in table A2 in Appendix II.

[4] For large samples ($n \geq 30$): 95% confidence limits = 1.96 SE
For small samples ($n < 30$): 95% confidence limits = $t_{0.05\,[n-1]}\ SE$

During a study of territoriality in dragonflies, the distance (in metres) at which resident males chased off approaching non-territory holders was measured. These data can be described using the mean, standard error and 95% confidence limits.

1. Record the data

The distances (in metres) measured for 15 residents are:
0.25 0.30 0.40 0.56 0.84 0.51 0.68 0.49 0.53 0.39 0.67 0.71 0.76 0.64 0.53

2. Calculate the mean value of each sample [1]

$$\bar{x} = \frac{\sum x}{n} = \frac{8.26}{15} = 0.551$$

3. Calculate the x^2 values and the sums of x and of x^2

x value	0.25	0.30	0.40	0.56	0.84	0.51	0.68	0.49	0.53	0.39	0.67	0.71	0.76	0.64	0.53
x^2 value	0.0625	0.09	0.16	0.3136	0.7056	0.2601	0.4624	0.2401	0.2809	0.1521	0.4489	0.5041	0.5776	0.4096	0.2809

The sum of the x values $(\Sigma x) = 8.26$
The sum of the x^2 values $(\Sigma x^2) = 4.9484$

Box 4.2 (continued)

4. Calculate the standard deviation of each sample [2]	$s = \sqrt{\dfrac{\sum x^2 - \dfrac{(\sum x)^2}{n}}{n-1}} = \sqrt{\dfrac{4.9484 - \dfrac{(8.260)^2}{15}}{15-1}} = \sqrt{\dfrac{0.3999}{14}} = 0.1690$
5. Calculate the standard error [3]	$SE = \dfrac{s}{\sqrt{n}} = \dfrac{0.1690}{\sqrt{15}} = 0.0436$
6. Calculate the 95% confidence limits [4]	Since n is less than 30, the 95% confidence limits = $t_{0.05\,[n-1]}\,SE$ Look up the t value in table A2 in Appendix II using $n - 1 = 15 - 1 = 14$ degrees of freedom $t_{0.05}$ at degrees of freedom of 14 = 2.145 95% confidence limits = $2.145 \times 0.0436 = 0.0936$
7. Interpret the results	The mean approach distance (plus/minus the standard error) by non-territory holders at which resident males are stimulated to chase them is 0.551 ± 0.0436 m. Alternatively, using the 95% confidence limits we can express this as 0.551 ± 0.0936 m. Further details of territorial behaviour in dragonflies are given by Miller (1995, NH 7). If you have several samples, you may wish to plot the mean and standard error (or 95% confidence limits) as a point chart or bar chart as in figs. 104 and 105 (Chapter 6).

Box 4.3 Medians and interquartile ranges

[1] Position of the median $= \frac{n+1}{2}$ Where: n is the size of the sample.

[2] Quartile 1 ($Q1$) is estimated as the nearest value to that indicated by a rank of $(n+1)/4$
Quartile 3 ($Q3$) is estimated as the nearest value to that indicated by a rank of $3 \times (n+1)/4$

An investigation into the impacts of leaf beetles on dock plants involved the assessment of leaf damage as an approximate percentage of the total leaf area. In order to summarise the impacts over several leaves, the median and interquartile range are calculated.

1. Record the data	For the 20 leaves examined the approximate percentage damage is as follows: 20 15 40 85 65 15 30 35 30 25 30 60 75 80 80 55 40 60 50 60
2. Put the data in ascending order	15 15 20 25 30 30 30 35 40 40 50 55 60 60 65 65 75 80 80 85
3. Calculate the position of the median and find the median value [1]	The position of the median = $(n + 1)/2 = (20 + 1)/2 = 10.5$ Thus, the median is at a position between the tenth (40) and eleventh (50) observation, i.e. is 45%
4. Calculate the position of each quartile [2]	The position of Q1 = $(n + 1)/4 = (20 + 1)/4 = 5.25$ The fifth value is 30% The position of Q3 = $3\times(n + 1)/4 = 3\times(20 + 1)/4 = 15.75$ The sixteenth value is 65%
5. Interpret the results	The median percentage leaf damage is 45%, with an interquartile range of 35% (Q1 = 30% and Q3 = 65%). The distribution is not quite symmetrical and has a bias towards the upper end of the range. Salt & Whittaker (1998, NH 26) give further details of insect feeding on dock plants. If you have several samples, you may wish to plot the median, quartiles and extreme values as a box and whisker plot as in fig. 106 (Chapter 6).

Estimating the reliability of the sample mean

Remember that any sample mean we calculate is an estimate of the true mean of the population. For example, if we count the numbers of animals under each of 50 logs in a woodland, we might use the mean number per log as an estimate of the mean number under all logs in that woodland. In such circumstances it would be useful to have some idea about how accurate our estimate of the true mean is. Once we have calculated the mean and the standard deviation we can calculate the reliability of our sample mean as an estimate of the true mean of the population. One measure of reliability is the standard error. Just as the standard deviation gives the range of the individual values that contribute to a particular sample mean, the standard error gives the range of sample means (if we took many samples) around the true population mean. It is in fact the standard deviation of sample means. Just as for the standard deviation, a range of the mean plus and minus one standard error encloses 68% of sample mean values. If we had only taken a single sample, then our single sample mean would fall within one standard error of the true mean 68% of the time. The standard error is calculated by dividing the standard deviation by the square root of the number of data points. The standard error should be expressed to one decimal place more than the mean.

If we take a large sample (more than 30) and a larger range around the mean of plus and minus 1.96 standard errors, our confidence in the estimate of the mean rises to 95%. The size of the range around the mean varies from 1.96 times the standard error where we have a large sample size (above 30) and rises with smaller sample sizes up to 12.706 times the standard error when we have only two data points. This range is termed the 95% confidence limits of the mean. For example, in a study of dragonfly territorial behaviour we can calculate the mean distance at which an approaching intruder stimulates the resident dragonfly to chase it (box 4.2). In our example this is 0.551 m. However, this is only an estimate of the true mean of the population of all possible dragonflies. The 95% confidence limits tell us that if we had repeated samples we would expect to find that the true mean lies between 0.4574 m and 0.6446 m (that is 0.551 ± 0.0936 m). Box 4.2 gives the formulae for the calculation of the standard error and 95% confidence limits for the distance at which territory holding dragonflies chase off intruders. Point or bar charts (figs. 104 and 105, Chapter 6) can be used to display means plus and minus standard errors or 95% confidence limits instead of standard deviations (pp. 46–47). Confidence limits can also be calculated for other descriptive statistics, such as population size estimates (p. 52).

Estimating population sizes and densities

It is relatively straightforward to calculate the density of static organisms such as plants and some seashore animals such as barnacles. The numbers of plants or animals are simply counted and divided by the area of the site. This gives the density in terms of the number per m^2, or, for organisms that are not very common, per hectare (that is, per 10,000 m^2). Total population size can be estimated by multiplying the density by the area involved. For example, if we have a density of 2.5 individuals per m^2, over an area of 20,000 m^2, the total population is 2.5 x 20,000 = 50,000. If the site is too large for the animals to be counted in the whole area, randomly placed quadrats can be used to sample the area (p. 4) and the number scaled up according to the sampling fraction (that is, the proportion of the site that has been sampled). For example, if 100 quadrats, 0.5 x 0.5 m in area, are used to randomly sample a site that is 20 x 25 m, the sampling fraction is the area sampled (100 x 0.5 x 0.5 m = 25 m^2) divided by the area of the site (20 x 25 m = 500 m^2). Here, we would have a sampling fraction of 0.05 (25/500) and the summed counts of all the quadrats would be multiplied by 500/25 to give the total population.

When animals are mobile, it is more difficult to be sure whether all individuals have been sampled. There are several techniques that can be used to estimate the population size, most of which require animals to be caught, marked, released and recaptured. Marking animals (technique, pp. 27–28) enables us to tell whether subsequent catches contain individuals that are new or recaptures. The majority of methods estimate the population by using the proportion of recaptures to the number caught on each occasion. The simplest such method is the Lincoln index (sometimes called the Peterson method). The sampling occasions are usually on successive days, but any time scale may be used as long as the marked animals have a chance to mingle thoroughly with the rest of the population. It is also possible to calculate how reliable our estimate of population size is, by calculating the standard error of the population estimate (box 4.4). The Lincoln index is biased when dealing with small samples and under such circumstances a correction factor may be used. Southwood & Henderson (2000) give details.

Population estimation methods rely on the assumption that there are no differences between marked and unmarked animals in behaviour, survival, or chance of recapture. Many of the other methods available are more sophisticated than the Lincoln index, and allow estimates to be made of survival as well as population size. Examples are the methods of Jolly, and of Fisher and Ford (Blower and others, 1981). However, in order to estimate survival, most techniques require several days of capture and recapture (in some cases at least a week).

A different type of method (variously called Craig's or Du Feu's method) is based on a single sampling occasion.

Here, animals are sampled and when caught, each is marked and released immediately. The subsequent captures are recorded as either new or recaptured animals and the total numbers of these two variables are used to calculate the population estimate. Unfortunately the formula for Du Feu's method is not simple and requires a computer in order to resolve it. Du Feu and others (1983) or Southwood & Henderson (2000) give further details.

A worked example of the Lincoln Index using weevils on nettles is given in box 4.4. Other situations in which these techniques may be useful might include, for example, population estimates of the pond skaters found on the surface of open water. The animals are caught and marked with a blob of flexible, waterproof paint (such as cellulose dope) on their backs, released and then later resampled. Guthrie (1989, NH 12) gives details of animals that live on the surface film. Population estimates may also be made of a subset of the animals in a particular area, such as the population of female butterflies laying eggs on cabbage plants (Kirk, 1992, NH 18), or of solitary wasps provisioning nests in logs (Yeo & Corbet, 1995, NH 3). Only those animals exhibiting the given behaviour are marked.

Box 4.4 Lincoln Index population estimation

[1] $\hat{P}_1 = \dfrac{M_1 \times M_2}{N_1}$

[2] $SE(\hat{P}_1) = \sqrt{\dfrac{M_1^{2} \times M_2 \times (M_2 - N_1)}{N_1^{3}}}$

Where: $\hat{P}_1$ is the population estimate on day 1;
$SE(\hat{P}_1)$ is the standard error of the population estimate;
M_1 is the number of animals captured, marked and released on day 1;
M_2 is the number of animals captured on day 2;
N_1 is the number of animals marked on day 1 and recaptured on day 2.

154 weevils were collected from patches of nettles. Having been marked and released on the first sampling occasion, the animals were then resampled on the next day and 282 weevils were collected, of which 81 had been captured and marked on day 1. The population size was then estimated using the Lincoln Index.

1. Record the data	The number of animals captured, marked and released on day 1 (M_1) = 154 The number of animals captured on day 2 (M_2) = 282 The number of animals marked on day 1 and captured on day 2 (N_1) = 81
2. Calculate the population estimate ($\hat{P}_1$) [1]	$\hat{P}_1 = \dfrac{M_1 \times M_2}{N_1} = \dfrac{154 \times 282}{81} = 536.1$
3. Calculate the standard error of the population estimate $SE(\hat{P}_1)$ [2]	$SE(\hat{P}_1) = \sqrt{\dfrac{M_1^{2} \times M_2 \times (M_2 - N_1)}{N_1^{3}}} = \sqrt{\dfrac{154^{2} \times 282 \times (282 - 81)}{81^{3}}} = \sqrt{2529.48} = 50.29$
4. Interpret the results	The population estimate for day 1 is 536.1 ± 50.29 weevils. Davis (1991, NH 1) and Morris (1991, NH 16) give further information about weevils on nettles.

Identifying clumped, regular and random distributions

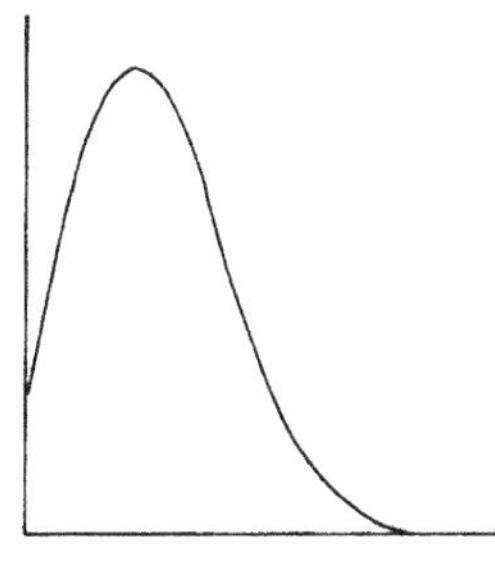

Fig 100. Example of the Poisson distribution when the mean is 2.

Where relatively rare items (such as the numbers of plants of certain species in quadrats, barnacles sparsely distributed on rocks, or plant galls on leaves) are distributed at random, a frequency histogram of quadrat counts may conform with a distribution called the Poisson distribution (fig. 100). A key feature of the Poisson distribution is that the variance is equal to the mean. For animals and plants recorded using quadrat sampling methods, this feature can be used to test whether the distribution is even, random or clumped.

Using this quadrat method, appropriately sized quadrats are chosen to sample the animals or plants in question, and a series of counts is made over the sampling area. For example, to sample limpets on a rock, we might place a 0.5 x 0.5 m quadrat in several (say, 20) randomly chosen positions on the rock, and count the limpets in each quadrat. The mean number of limpets per quadrat and the variance (the square of the standard deviation, box 4.2) are calculated. If the variance:mean ratio (the variance divided by the mean) is about 1, that is, the variance and the mean are nearly equal, then the limpets are distributed randomly. This ratio is called the index of dispersion. However, if the variance is much smaller than the mean (index of dispersion < 1), then there is little variation in the numbers of limpets found in a quadrat and they show an even, or regular, distribution. On the other hand, if the variance is much larger than the mean (index of dispersion > 1), there is more variation in the numbers of limpets found per quadrat; some quadrats have many individuals and others have relatively few. This distribution is clumped. To be more certain whether or not an index of dispersion of nearly one really represents a non-random distribution, we use a statistical inferential test (described on p. 72).

Box 4.5 shows the how distribution patterns of solitary wasp nests are identified using quadrat sampling. Many other questions might be addressed using quadrat methods to assess the distribution of organisms. For example, are millipedes aggregated under rotting logs (Wheater & Read, 1996, NH 22)? By sampling animals under a large number of logs, the variance:mean ratio can be calculated and interpreted using this technique. Does egg laying by small white butterflies on cabbages result in clumping of eggs (Kirk, 1992, NH 18)? Cabbage leaves can be sampled to see how many eggs have been laid on each and the variance:mean ratio examined.

The index of dispersion may depend on the scale of sampling. For example, if sampling is confined to the most popular area, the distribution of deck chairs on a crowded beach will appear regular, with each chair placed a comfortable distance from the next. But if sampling is extended to include less favoured areas of beach, the distribution will appear clumped, with deck chairs clustered

Box 4.5 Examining distributions using quadrat methods

[1] Index of dispersion = $\frac{\text{variance}}{\text{mean}}$ Where: the variance (s^2) is the square of the standard deviation (s)

In a survey of a colonially nesting solitary wasp species, an area of 4 square metres containing 112 nests was examined to see whether the nests were distributed randomly or were either aggregated or evenly distributed. A series of 25 randomly placed small quadrats (0.1 m x 0.1 m) was used to sample the nests.

1. Record the data	Numbers of nests recorded per quadrat are 1 3 3 1 0 3 2 1 4 0 0 2 0 0 2 0 0 1 1 1 2 0 1 0 0
2. Calculate the mean number of nests per quadrat (see box 4.2)	The mean number of nests per quadrat = 1.12
3. Calculate the standard deviation (see box 4.2) and variance (the square of the standard deviation) of nests	The standard deviation of the nests (s) = 1.201 The variance (s^2) = 1.201^2 = 1.443
4. Calculate the index of dispersion [1]	Index of dispersion = $\frac{\text{variance}}{\text{mean}} = \frac{1.443}{1.12} = 1.289$
5. Interpret the results	Since the index of dispersion is above 1 there is a tendency towards a clumped distribution. An additional test is required to confirm whether this is a significantly clumped distribution (see Chapter 5). Yeo & Corbet (1995, NH 3) give more information about the ecology of solitary wasps.

in the most popular area. Greig-Smith (1983) gives more information about the index of dispersion.

An alternative technique employs a plotless sampling method, in which the calculations are based on the distances between nearest neighbours. For example, to examine the distribution of trees in a woodland, we begin with a randomly chosen tree and measure the distance to its nearest neighbouring tree. By repeating this several times on randomly chosen trees, we obtain a series of distance measurements. We calculate the density of trees in the area and the mean distance between nearest neighbours and compare the latter with the theoretical distance that we would expect if the trees were randomly distributed, bearing in mind that we know the density. The density and distance should be expressed in corresponding units; if the distances are measured in metres, then the density must be calculated as the number per square metre. If the ratio between the observed mean distance between nearest neighbours and that expected from a random distribution (the index of aggregation) is greater than one, then the trees are more evenly distributed than random. On the other hand, if the ratio is less than one then there is a tendency for the distribution to be clumped (box 4.6). A simple test enables us to identify whether the distribution is significantly clumped, random or regular (p. 75). With small samples (less than 100), this method is reasonably robust only if the plot shape is not too elongated (that is, a square or circle rather than a long

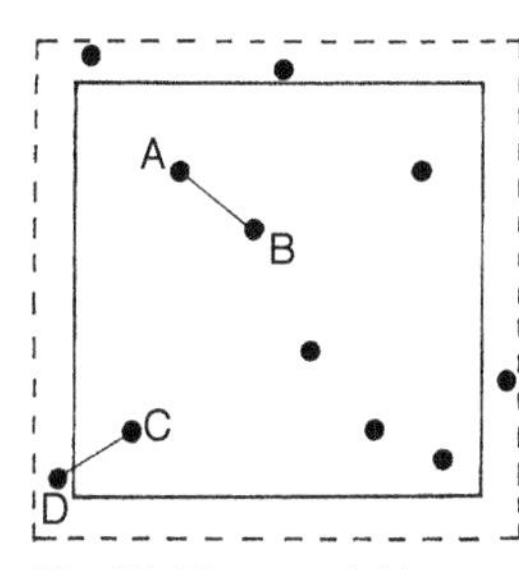

Fig. 101. Nearest neighbour sampling showing boundary zone (A, B and C lie within the sampling area, but, the nearest neighbour to C is D which lies within the boundary zone and is used only for the nearest neighbour measurements).

rectangle) and if a boundary zone is included within the plot layout such that nearest neighbours of individuals within the study plot may lie outside the plot but within the boundary zone (fig. 101). Krebs (1999) gives more information about this technique.

A worked example of the nearest neighbour method of examining distribution patterns using lugworm burrow distribution is given in box 4.6. Other situations in which nearest neighbour methods are useful in assessing the distribution of organisms might include studies of the distribution of large, widely spaced items for which quadrat methods would be impracticable. For example, one might investigate whether ants' nests in meadows (Skinner & Allen, 1996, NH 24) are significantly clumped, evenly or randomly distributed. The nearest neighbour technique can also be used at a very small scale. For example, by examining the distances between greenflies feeding on leaves one can identify whether they are clumped in particular areas. If it is difficult to measure the distances between individuals *in situ*, leaves can be photographed and measurements taken from the photographs later. A ruler or other scale object should be included in the photograph so that all the measurements can be standardised. A classic study by Kennedy & Crawley (1967) used the nearest neighbour method to demonstrate the nature and causes of 'spaced-out gregariousness' in sycamore aphids.

Examining community diversity

In ecological investigations we sometimes need to quantify the species richness or diversity of a community of animals or plants. On its own, a list of the numbers and types of animals found in a survey of, say, a patch of nettles, or thistles, or seaweed, may not convey much about diversity. We could express community richness or diversity in terms of species richness: the number of species (or families, or orders) per habitat unit (patch). But this simple measure of species numbers takes no account of the fact that some species are rare whilst others are common. There are several diversity indices that reflect the relative proportions of different species in the community. Most of these diversity indices increase in value when there are more species and when the relative proportions of the different species in a site are reasonably uniform. The full range of diversity indices is beyond the scope of this book, but Magurran (1988) gives further details. Several simple indices in common use, which are based on the relative abundances of the species found in a site, are shown in box 4.7. Some, such as the Shannon index, are biased towards rare species, and others, such as the Simpson index, towards common ones. There has been some debate about the value of many of these indices. One of the simplest to calculate, and one with more support than most, is the Berger–Parker index. This is based on the ratio of the number of individuals of the most abundant species to

Box 4.6 Examining distributions using nearest neighbour methods

[1] $r_d = \frac{\Sigma r_i}{n}$

[2] $d = \frac{\text{number within study area}}{\text{area of study area}}$

[3] $r_e = \frac{1}{2\sqrt{d}}$ [4] $R = \frac{r_d}{r_e}$

Where: r_d is the mean distance between nearest neighbours;
r_i is the measured distance to the nearest neighbour for a particular individual;
n is the number of individuals in the study area;
d is the density of organisms (in the square of the units used for the nearest neighbour measurements);
r_e is the expected distance between nearest neighbours if the distribution is random;
R is the index of aggregation.

A survey of a seashore revealed a large number of burrows formed by lugworms (1050 burrows in a 5 m x 5 m plot). Within this plot 20 individual burrows were identified randomly and the distances to their nearest neighbours were measured. A boundary zone was incorporated within which burrows were not sampled, but which could contain nearest neighbours for those being sampled.

1. Record the data	The distances (in millimetres) between nearest neighbours are as follows: 65 70 50 55 45 45 60 60 80 75 60 40 65 75 55 45 60 50 55 65
2. Calculate the mean distance between nearest neighbours [1]	$r_d = \frac{\Sigma r_i}{n}$ $= \frac{65+70+50+55+45+45+60+60+80+75+60+40+65+75+55+45+60+50+55+}{20}$ $= \frac{1175}{20} = 58.75$
3. Calculate the density of the organisms [2]	$d = \frac{\text{number within study area}}{\text{area of study area}} = \frac{1050}{25} = 42$ burrows per square metre Since the nearest neighbour measurements were taken in millimetres, the density of burrows per square metre must be converted to the number per square millimetre by dividing by 1 000 000 (there being 1000 x 1000 square millimetres in a square metre), hence the density is 0.000042 per square millimetre.
4. Calculate the expected distance between nearest neighbours [3]	$r_e = \frac{1}{2\sqrt{d}} = \frac{1}{2\sqrt{0.000042}} = 77.1517$
5. Calculate the index of aggregation [4]	Index of aggregation (R) $= \frac{r_d}{r_e} = \frac{58.75}{77.1517} = 0.7615$
6. Interpret the results	Since the index of aggregation is below 1 there is a tendency towards a clumped distribution. An additional test is required to confirm whether this is a significantly clumped distribution (see Chapter 5). Hayward (1994, NH 21) gives further information about the distribution of lugworm burrows.

the total number of individuals at the site, and is also shown in box 4.7. Since the original formula for the Simpson index falls with increasing diversity, several alternative formulae have been proposed to alter this so that the index gets larger as the diversity rises. This has led to some confusion with different authors using slightly different calculations. We have used the simplest of these in box 4.7. Magurran (1988) describes several techniques that allow statistical comparisons between the diversity indices of different sites.

The evenness, or equitability, of community structure may also be calculated as an index lying between zero (heavily dominated by a single species, with all others only rarely found) and one (where all species are found in equal numbers). There are many different types of evenness index. Those in most common use are based on the ratio between a calculated diversity index and its maximum possible value, since most diversity indices are maximal (for any particular number of species) when the individuals are evenly distributed among all the species present. One of the most commonly employed of these indices is based on the Shannon diversity index (Krebs, 1999) and is given in box 4.7.

Although it is common to examine site diversity on the basis of species, higher levels of taxonomic grouping, such as families, are sometimes used. However, there are problems in the use of such coarse taxonomic levels. Species-based indices are more useful because a species is usually considered a real biological entity, whereas higher taxa, such as families, vary with the opinion of the taxonomist who created them: one taxonomist might classify a group into a few families, each with many species, whereas another might prefer to establish more numerous, smaller families. It is not always necessary to name the species (or families) of animals (or plants) in order to assess diversity. The catch can be sorted into types that look different from one another (some people dignify these batches of look-alikes with the term 'morphospecies') and the index of diversity can be calculated in terms of the numbers of individuals of each type. A common error is to assign conspecific males and females to separate morphospecies. To avoid this one can adopt an *ad hoc* convention, such as scoring only males of chironomid midges caught in water trap samples.

Taxonomic classification is not the only basis for describing community structure using diversity indices; instead we could use ecological equivalents, such as the proportion of predators, herbivores, decomposers and mixed feeders in habitats such as patches of nettles, or on seaweed, or under stones. These feeding types may be further subdivided. For example, insect herbivores may be divided into groups including chewers, leaf miners, gall formers and sap suckers, whilst predators may be divided into hunters and trappers. Diversity indices can then be used to express the richness of the community based on feeding type.

Box 4.7 illustrates the calculations of several diversity indices using the example of the settlement of bryozoans on seaweeds. Other situations in which diversity and evenness indices are useful include, for example, a study of the diversity of the invertebrate community found on a species of plant. Particularly suitable plants are nettles (Davis, 1991, NH 1), thistles (Redfern, 1995, NH 4), cabbages (Kirk, 1992, NH 18), Sphagnum (Hingley, 1993, NH 20), docks (Salt & Whittaker, 1998, NH 26) or cherry trees (Leather & Bland, 1999, NH 27).

Box 4.7 Diversity and evenness indices

[1] The Shannon Index $(H') = -[\Sigma p_i \ln p_i]$

[2] The Simpson Index $(D) = \Sigma p_i^2$

[3] The Berger–Parker Index $(d) = \frac{N_{max}}{N}$

[4] The Evenness Index $(J') = \frac{H'}{\ln S}$

Where: p_i is the proportional abundance of each species at the site;
ln is the natural logarithm;
N_{max} is the number of individuals in the most abundant species at the site;
N is the total number of individuals at the site;
S is the number of species found at the site.

Note that the Simpson Index is often expressed as $1 - D$ while the Berger–Parker Index is usually expressed as its reciprocal $(1/d)$ in order to produce indices that rise with increasing diversity.

Studies of the seaweeds *Fucus serratus, F. spiralis* and *F. vesiculosus* reveal different settlement patterns of four species of bryozoans. The diversity and evenness of these can be calculated for each species using a variety of indices.

1. Record the data in terms of the number of each species of bryozoan on each species of seaweed. Note that not all species must be found at all sites.

Bryozoan species	***F. serratus***	***F. spiralis***	***F. vesiculosus***
Alcyonidium hirsutum	35	19	6
A. gelatinosum	33	13	9
Flustrellidra hispida	37	23	15
Celleporella hyalina	9	6	1
TOTAL	114	61	31

2. Calculate the proportions of each species per site or plant, that is, the number of each species found divided by the total for that site or plant e.g. for *A. hirsutum* on *F. serratus*, $\frac{35}{114} = 0.3070$

Bryozoan species	***F. serratus***	***F. spiralis***	***F. vesiculosus***
Alcyonidium hirsutum	0.3070	0.3115	0.1935
A. gelatinosum	0.2895	0.2131	0.2903
Flustrellidra hispida	0.3246	0.3770	0.4839
Celleporella hyalina	0.0789	0.0984	0.0323

Box 4.7 (continued)

3. Calculate the indices

For example, for *F. serratus*:

[1] Shannon (H')

$$= -[(0.3070 \times \ln[0.3070]) + (0.2895 \times \ln[0.2895]) + (0.3246 \times \ln[0.3246]) + (0.0789 \times \ln[0.0789])]$$

$$= -[(0.3070 \times -1.1809) + (0.2895 \times -1.2397) + (0.3246 \times -1.1253) + (0.0789 \times -2.5390)] = \mathbf{1.2871}$$

[2] Simpson (D)

$$= (0.3070)^2 + (0.2895)^2 + (0.3246)^2 + (0.0789)^2$$

$$= 0.0943 + 0.0838 + 0.1053 + 0.0062 = \mathbf{0.2896}$$

Simpson ($1 - D$)

$$= 1 - 0.2896 = \mathbf{0.7104}$$

[3] Berger-Parker (d)

$$= \frac{37}{114} = \mathbf{0.3246}$$

Berger-Parker ($1/d$)

$$= \frac{1}{0.3246} = \mathbf{3.0811}$$

[4] Evenness (J')

$$= \frac{1.2871}{\ln(4)} = \frac{1.2871}{1.3863} = \mathbf{0.9284}$$

4. Interpret the results

Diversity Indices	***F. serratus***	***F. spiralis***	***F. vesiculosus***
[1] Shannon (H')	1.2871	1.2887	1.1389
[2] Simpson (1 – D)	0.7104	0.7057	0.6431
[3] Berger-Parker ($1/d$)	3.0811	2.6522	2.0667
[4] Evenness (J')	0.9284	0.9296	0.8216

Irrespective of the diversity index used, the least diverse community of bryozoans appears to be on *F. vesiculosus*, with *F. serratus* and *F. spiralis* having the highest diversities. Hayward (1988, NH 9) gives further details about bryozoans on seaweeds.

Examining similarities between communities

Similarity indices allow us to quantify similarities between different sites on the basis of the species held in common. There are basically two types of method. In one, the species present are compared between two sites to identify how many species are held in common, and how many are unique to each site. Some variations on this method

Box 4.8 Percentage similarity

[1] $P_{A,B} = \Sigma \text{ minimum } (p_{Ai}, p_{Bi})$
i.e. the percentage similarity between any two sites = the sum of the lowest percentage of each pair of sites

Where: $P_{A,B}$ is the percentage similarity between sites A and B;
p_{Ai} is the percentage of species i in site A;
p_{Bi} is the percentage of species i in site B.

Investigations of fallen logs revealed that a number of ground beetles were living beneath them. Twenty five logs from each of four sites were compared to examine the similarity of the ground beetles found in each habitat type.

1. Record the data as the number of individuals in each species. Note that not all species need to be present in all sites

Group of animals	Woodlands	Wood edges	Hedgerows	Gardens
Pterostichus madidus	7	5	12	0
Nebria brevicollis	4	3	9	2
Nebria salina	0	1	1	6
Notiophilus biguttatus	6	0	4	1
Loricera pilicornis	2	0	1	0
Calathus melanocephalus	1	1	3	1
TOTAL	20	10	30	10

2. Calculate the percentage that each species forms of the total for that site

For *P. madidus* in woodland

(7 × 100) / 20 = 35%

Group of animals	Woodlands	Wood edges	Hedgerows	Gardens
Pterostichus madidus	35%	50%	40%	0%
Nebria brevicollis	20%	30%	30%	20%
Nebria salina	0%	10%	3.3%	60%
Notiophilus biguttatus	30%	0%	13.3%	10%
Loricera pilicornis	10%	0%	3.3%	0%
Calathus melanocephalus	5%	10%	10%	10%

3. Calculate the percentage similarity between each pair of sites (i.e. the sum of the lowest percentage in each pair – if the percentages are the same, then use that value) [1]

First comparing woodlands and wood edges, *P. madidus* makes up 35% of woodlands and 50% of wood edges, so we take the lower value (35%) and add it to the lowest value for *N. brevicollis* (20%), then *N. salina* (0%), *N. biguttatus* (0%), *L. pilicornis* (0%) and, finally, *C. melanocephalus* (5%), giving the percentage similarity between woodlands and wood edges as:
35 + 20 + 0 + 0 + 0 + 5 = 60%
That is, they are 60% similar. This is repeated for each site pair and a matrix produced:

	Woodlands	Wood edges	Hedgerows
Wood edges	60%		
Hedgerows	76.6%	**83.3%**	
Gardens	35%	40%	43.3%

At 83.3%, Wood edges and Hedgerows have the greatest percentage similarity

4. Select the two sites with the highest percentage similarity, combine these in terms of actual numbers of animals and recalculate the percentages

Group of animals	Wood edges + Hedgerows	Percentages
Pterostichus madidus	17	42.5%
Nebria brevicollis	12	30%
Nebria salina	2	5%
Notiophilus biguttatus	4	10%
Loricera pilicornis	1	2.5%
Calathus melanocephalus	4	10%
TOTAL	40	

Box 4.8 (continued)			
5. Recalculate the percentage similarity between each pair of sites	Wood edges + hedgerows Gardens	72.5% 35% 45% Woodlands Wood edges + Hedgerows	**At 72.5%, Woodlands and Wood edges + Hedgerows have the greatest percentage similarity**
6. Select the pair with the highest percentage similarity, combine the sites in terms of actual numbers of animals and recalculate the percentages	**Group of animals**	**Woodlands + Wood edges + Hedgerows**	**Percentages**
	Pterostichus madidus	24	40%
	Nebria brevicollis	16	26.7%
	Nebria salina	2	3.3%
	Notiophilus biguttatus	10	16.7%
	Loricera pilicornis	3	5%
	Calathus melanocephalus	5	8.3%
	TOTAL	60	
7. Recalculate the percentage similarity between each pair of sites		41.6% Gardens	**Woodlands + Wood edges + Hedgerows and Gardens have a percentage similarity of 41.6%**
8. Interpret the results. It may be useful to draw the relationships between the sites using a dendrogram to show the sites linked at the calculated similarity levels (note that the scale starts with 100% at the bottom)	In terms of the percentage composition of their ground beetle fauna, dead wood in wood edges and hedgerows are the most similar, followed by logs found in woodlands. Dead wood in gardens are the least similar in ground beetle percentage composition to logs found in the other habitats. Forsythe (2000, NH 8) gives further information about ground beetles and Wheater & Read (1996, NH 22) discuss animals under logs.	Percentage similarity: 40%, 50%, 60%, 70%, 80%, 90% Sites: Wood edges, Hedgerows, Woodlands, Gardens	

also take into account double absences, that is species that were found during the study, but were not present in either of the two sites being compared. But these presence-and-absence methods do not make full use of all the information available. They give the same weight to rare species as to common ones because they do not take into account the numbers of individuals of each species. Krebs (1999) and Southwood & Henderson (2000) give further details of presence-and-absence methods.

A second approach does take into account the numbers of individuals of each species found in each site, weighting the species in terms of their abundance. One such method compares the percentages of each species in the two sites and is called the percentage similarity or the Renkonen index (box 4.8). If there are more than two sites, all combinations of pairs of sites are compared, one pair at a time. This process is done repeatedly to generate a display of the similarities of the sites using a dendrogram or tree diagram.

Although it is common to compare sites on the basis of species, higher levels of taxonomic grouping, such as families, are sometimes used in similarity analysis, but there are problems in the use of such coarse taxonomic levels. Species-based indices are more useful because, as we have argued (p. 59) a species is usually considered a real biological entity, whereas higher taxa, such as families, vary with the opinion of the taxonomist who created them.

Box 4.8 shows the calculation of percentage similarity using the example of ground beetle communities under fallen logs in a variety of habitat types. Similarly, the similarity of component species of mayfly larvae could be examined in several streams that differ in altitude or in the amount of organic pollution present (Harker, 1989, NH 13). Other situations in which similarity indices are useful might include examination of the similarity in food preference between several species of predatory insects, such as hoverfly, dragonfly or ladybird larvae, ground beetle adults, or worker ants. This may be measured as the percentages of each of several prey items taken in laboratory trials. The diets of hoverflies, dragonflies, ladybirds, ground beetles and ants are discussed by Gilbert (1993, NH 5), Miller (1995, NH 7), Majerus & Kearns (1989, NH 10), Forsythe (2000, NH 8) and Skinner & Allen (1996, NH 24) respectively. The similarity of the invertebrate communities found on different, but closely related, species of plants may also be investigated in this way. For example, a comparison could be made of the invertebrates found on several species of thistles (Redfern, 1995, NH 4). Similar observations could be made on invertebrates on other plants such as brassicas (Kirk, 1992, NH 18), Sphagnum (Hingley, 1993, NH 20), docks (Salt & Whittaker, 1998, NH 26), cherry trees (Leather & Bland, 1999, NH 27) or seaweeds (Hayward, 1988, NH 9), or in the species found in collections made at different times of year.

5 Statistical testing

Inferential statistical tests provide us with objective answers to questions about differences between samples, relationships between variables and associations between frequencies of categorical variables. Statistical tests help us to test a null hypothesis. A null hypothesis proposes that there is no difference between samples (or no relationship between variables, or no association between frequency distributions). When testing for a difference between two samples, the statistical test asks how likely it is that the two samples could have come from the same population. If that probability is high, they are very likely to have come from the same population, and we cannot reasonably claim that they are different. If the probability is very low, then the two samples are unlikely to have come from the same population and we can reject the null hypothesis and accept the alternative hypothesis that there is a significant difference between the two samples. Similar null hypotheses can be set up and tested for relationship and association questions.

Probabilities range from zero (no probability of an event taking place) to one (absolute certainty of an event taking place). For example, if you toss a coin, there is a one in two chance of its landing with the head side uppermost. This could also be expressed as a probability of 50% or 0.5. By convention, the threshold of statistical significance is usually set at a probability of 5%, or 0.05. If the probability (*P*) that two samples came from the same population is less than 5% (that is, if $P < 0.05$), we reject the null hypothesis and accept an alternative hypothesis, claiming that there is a significant difference between the samples. If the probability is 5% or more, we cannot make that claim. But nor can we claim that they are the same. There might be a small difference between them that could be revealed by a more exhaustive sampling programme. If we fail to find a difference we cannot say 'there is no difference between the two samples'. Instead, we can say 'there is no significant difference', allowing for the possibility that a study with a larger sample size might have found one. Care should be taken when interpreting probabilities close to the critical value ($P = 0.05$) based on small sample sizes, since smaller sample sizes may be less representative of the population as a whole. There are other possible reasons why *P* may be greater than 0.05. The null hypothesis might really be true (that is, there might really be no difference between the samples). Alternatively, the methods used to measure the data points might be so inaccurate that the noise in the data obscures any possible difference.

For those new to statistical ideas, it may take a while to get used to working in this apparently negative way, testing only null hypotheses. But it is important to think this through, because the explicit formulation of the null hypothesis is an essential first step in designing a survey or experiment and in analysing the data, and it affects the way you express your findings. Many people become worried if they find a non-significant result, thinking that they have not obtained the correct answer, or that the experiment or survey 'didn't work'. On the contrary, such findings are frequently of importance since they may allow us to reject assumptions and move arguments on, enabling us to concentrate our attention on other factors that might have more immediate influences. It is this process of acquiring knowledge by small steps that progresses much of the scientific knowledge we have today.

The basic steps in the testing procedure are alike whichever test is to be used (fig. 102). The first step is to decide on the question to be asked (whether we are testing for a difference between samples, or a relationship between variables, or an association between the frequencies of categorical variables) and to formulate the null hypothesis precisely. The second step involves choosing a statistical test appropriate for the data and question being tested (Keys A, B

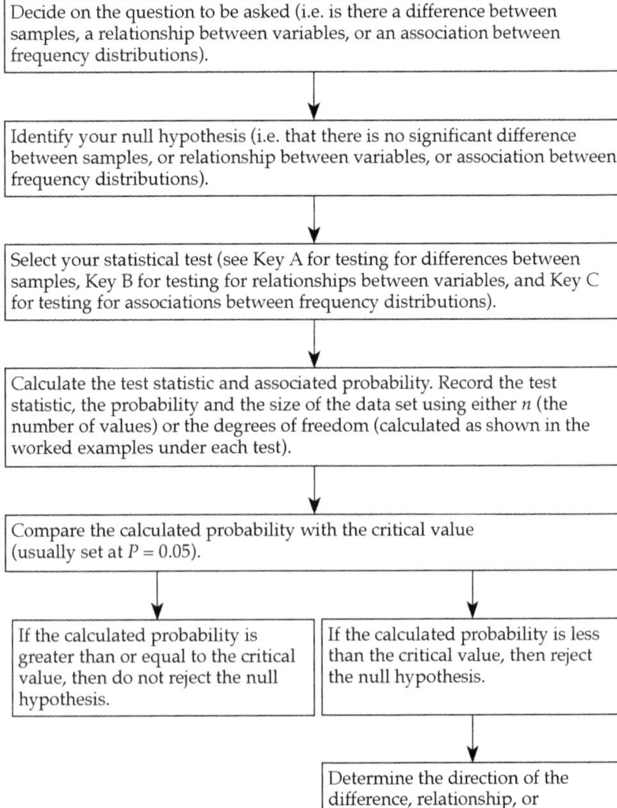

Fig. 102. Sequence of steps in calculating statistical tests.

and C, pp. 69, 84 and 89 respectively). The test statistic is then calculated using the formulae given in the worked examples in the following sections, and the answer recorded. Using the test statistic and the size of the data set, tables (or a computer) are then used to identify the probability of getting the results obtained if the null hypothesis is correct. The value used to indicate the size of the data set may be the actual number of data points (given the symbol n), or a number called the degrees of freedom (df) which is related to the size of the data set but adjusted slightly because we are making estimations from a sample (see Wheater & Cook, 2000, for further explanation). The number of degrees of freedom is calculated using a simple formula which is given when each test is discussed in the following sections. Examples of how to look up probabilities using degrees of freedom and number of data points are given for two tests in tables 3 and 4. The test statistic, the degrees of freedom (or actual number of data points, depending on the test) and the probability should always be recorded and reported in any write up of your work. Two decimal places are usually sufficient for recording test statistics, while probabilities are often taken to three decimal places. However, if the test statistic is a large number (over 100) there is usually no need to include any decimal places.

Having arrived at a test statistic and looked up the P value, if the probability is low (below the critical value of 0.05), we need to examine the data to find out the nature of the difference, relationship or association. For example, we may have shown a significant difference between samples, but now we need to ask which is larger, or has more individuals or species.

The formulae are given here for completeness, but it is often simpler and quicker to use a computer program such as Excel, Minitab or SPSS to calculate these test statistics. Dytham (2003) gives a brief introduction to the use of these programs for the statistical analysis of biological data. However commercial statistical programs may be expensive and many naturalists do not have access to them. An easy-to-use Excel workbook is available from the authors (address p. 100) that illustrates the calculations using the data from the worked examples in this book and can be used to analyse your own data.

Statistical tests are of two main types: parametric tests and nonparametric tests. Parametric tests are appropriate if the data consist of measurements (rather than ranked or categorical data). In addition, the measurements should be at least approximately normally distributed, that is, with a symmetrical frequency distribution and approximately 95% of the data points occurring within two standard deviations either side of the mean ($\bar{x}$ {termed '*x* bar'} $\pm 2s$) (pp. 44–49). Nonparametric tests, on the other hand, are appropriate when data are ranked, the frequency distribution is not normal, or both. Since nonparametric tests give slightly more conservative answers than the equivalent parametric tests, parametric tests are preferable when the data are appropriate. If the data are not normally distributed they can sometimes be manipulated mathematically before analysis to bring them closer to a normal distribution. This process, called transformation, is described in Wheater & Cook (2000). With very small samples it is difficult to tell whether the data points are normally distributed or not. When in doubt, it is safer to use nonparametric tests. The following sections include examples of parametric tests and their nonparametric equivalents.

Table 3. *Selected values of* t *to illustrate how to find the critical* t *value for the example in box 5.1.* The *t* value calculated (2.31) is compared with that found at the appropriate degrees of freedom (see the shaded row). Calculated values of *t* greater than the table values are significant. A more comprehensive table of *t* values is given in table A2 in Appendix II.

df ($n_1 + n_2 - 2$)	*t*	
	$P = 0.05$	$P = 0.01$
14	2.145	2.977
15	2.131	2.947
16	2.120	2.921
17	2.110	2.898
18	2.101	2.878

Table 4. *Selected values of* U *for the Mann–Whitney* U *test to illustrate how to find the critical* U *value for the example in box 5.4.* The *U* value calculated (25) is compared with that found at the appropriate *n* values (see the shaded areas). For significance, calculated *U* values must be smaller than the table value. A more comprehensive table of *U* values is given in Table A3 in Appendix II. The upper table values (in bold) in each row equate to $P = 0.05$, while the lower values are for $P = 0.01$.

Smaller *n* value	Larger *n* value: 7	8	9	10	11	12
7	**8**	**10**	**12**	**14**	**16**	**18**
	4	6	7	9	10	12
8		**13**	**15**	**17**	**19**	**22**
		7	9	11	13	15
9			**17**	**20**	**23**	**26**
			11	13	16	18
10				**23**	**26**	**29**
				16	18	21
11					**30**	**33**
					21	24
12						**37**
						27

Choosing a statistical test

The statistical test chosen depends firstly on the question being asked. Three major questions are usually asked:

Is there a significant difference between two (or more) samples? Key A (below)

Is there a significant relationship between two variables? Key B (p. 84)

Is there a significant association between the frequency distributions of two variables comprising categorical data? Key C (p. 89)

Testing for differences between samples

When choosing a test for differences between two or more samples, we need to know whether there are more than two samples and whether the data are matched or not (p. 5). This section only discusses comparisons involving two samples. For comparisons involving more than two samples, see Wheater & Cook (2000). When planning your study take care to avoid complex designs. For example, if different types of water trap are being compared on the basis of their colour (white compared with yellow) and the influence of a potential attractant, it might seem sensible to combine the treatments and have white traps with and without attractants as well as yellow traps with and without attractants. However, such a design requires quite complex analysis since there are two types of treatment (colour and presence or absence of attractant), and these may interact in some way. Wheater & Cook (2000) discuss such designs. It is usually better to keep the situation simple in the first instance and deal with each treatment type separately.

Key A Statistical tests for differences

1 Testing for a difference between two samples 2

– Testing for differences between more than two samples Analysis of variance
See p. 81 for a brief description and Wheater & Cook (2000) for more information

2 Data are unmatched 3

– Data are matched 5

3 Data are measurements (that is, interval or ratio data) 4

– Data are measured on a ranked scale Mann–Whitney *U* test (p. 75)

Box 5.1 *t* test

[1] $$t = \frac{|\bar{x}_1 - \bar{x}_2|}{\sqrt{\frac{(n_1 - 1)s_1^2 + (n_2 - 1)s_2^2}{(n_1 + n_2 - 2)} \times \frac{n_1 + n_2}{n_1 n_2}}}$$

[2] $$\text{df} = n_1 + n_2 - 2$$

Where: $|\bar{x}_1 - \bar{x}_2|$ is the absolute difference (that is, converting any negative values to positive) between the mean of the first sample ($\bar{x}_1$) and the mean of the second sample ($\bar{x}_2$);
s_1^2 and s_2^2 are the variances of samples 1 and 2;
n_1 and n_2 are the sample sizes of samples 1 and 2;
df are the degrees of freedom.

A survey investigated the duration of bumblebee visits to flowers on plants that formed part of ten clumps and those on nine plants that were isolated from other similar plants. Comparisons between the handling times on the two different types of plant may be made using a *t* test.

1. Record the data	Time (in seconds) spent on flowers on plants in a clump 5 4 3 3 4 4 5 6 4 5 Time (in seconds) spent on flowers on isolated plants 4 5 7 6 3 7 5 6 8
2. Calculate the mean value of each sample (see box 4.2)	Mean time on flowers on clumped plants ($\bar{x}_c$) = 4.3 seconds Mean time on flowers on isolated plants ($\bar{x}_i$) = 5.67 seconds
3. Calculate the standard deviation of each sample (see box 4.2)	Standard deviation of time on flowers on clumped plants (s_c) = 0.949 seconds Standard deviation of time on flowers on isolated plants (s_i) = 1.581 seconds
4. Calculate the number of data points in each sample	Number of data points for flowers on clumped plants (n_c) = 10 Number of data points for flowers on isolated plants (n_i) = 9

Box 5.1 (continued)	
5. Calculate the *t* statistic [1]	$t = \dfrac{\lvert \bar{x}_c - \bar{x}_i \rvert}{\sqrt{\dfrac{(n_c - 1)s_c^2 + (n_i - 1)s_i^2}{(n_c + n_i - 2)} \times \dfrac{n_c + n_i}{n_c n_i}}}$ $= \dfrac{\lvert 4.30 - 5.67 \rvert}{\sqrt{\dfrac{(10-1)0.949^2 + (9-1)1.581^2}{(10+9-2)} \times \dfrac{10+9}{10 \times 9}}} = 2.314$
6. Calculate the degrees of freedom [2]	$df = n_c + n_i - 2 = 10 + 9 - 2 = 17$
7. Look up the probability (*P*). See table A2 in Appendix II	For $t = 2.31$ with $df = 17$, $P < 0.05$ (computer programs give an exact value of $P = 0.033$)
8. Do not reject the null hypothesis if the calculated probability is greater than or equal to the critical level OR reject the null hypothesis if the calculated probability is less than the critical level and examine the means to determine where the difference lies	Here *P* is less than the critical level of 0.05. Therefore we reject the null hypothesis and, since the mean duration of bumblebee visits on clumped plants is less than that on isolated plants (see step 2), we can say that: "Bumblebees spend significantly less time per flower visit on clumped plants than on isolated plants ($t = 2.31$, $df = 9$, $P < 0.05$)." It may be that when there are fewer plants nearby, it is worth spending extra time probing the flowers, whereas when there are potential alternative sources of food close by, handling can be swifter. Prŷs-Jones & Corbet (1991, NH 6) give more information on bumblebee ecology and behaviour.

4 Data approximate to a normal distribution *t* test (below)
– Data do not approximate to a normal distribution
Mann–Whitney *U* test (p. 75)

5 Data are measurements (that is, interval or ratio data) 6
– Data are measured on a ranked scale
Wilcoxon matched pairs test (p. 81)

6 Differences between pairs of data approximate to a normal distribution Paired *t* test (p. 80)
– Differences between pairs of data do not approximate to a normal distribution Wilcoxon matched pairs test (p. 81)

t test

The *t* test is very commonly employed, and the calculation of the test statistic (t) is based on examining the difference between the means of two samples, while taking into account the variation (standard deviation) of each sample and the number of data points sampled. If a significant difference is obtained, it is necessary to examine the mean values of the two samples to interpret where the difference lies. Box 5.1 uses the example of foraging time in bumblebees to show the workings of this test. Here the mean time spent handling flowers on plants found in a clump is compared with the mean time spent handling flowers on isolated plants. Table 3 (p. 68) illustrates how to look up a *t* statistic to obtain the probability for this example.

Other situations in which the *t* test may be useful include comparisons of counts of common items, such as the numbers of insects found on two different species of cherry trees, or the numbers of blowflies found on carrion in the shade as opposed to carrion in direct sunlight. Leather & Bland (1999, NH 27) and Erzinçlioğlu (1996, NH 23) give details of insects on cherry trees and blowfly ecology respectively. Care should be taken when counts are of rare items since these may be skewed and thus not conform to a normal distribution (see p. 44). The *t* test is also useful when comparing two types of habitats on the basis of measurements such as the length or mass of individuals of a species found there, or of some characteristic of the physical or chemical nature of the habitats, such as soil temperature or pH.

The *t* test can also be used as a one-sample test to examine the significance of a comparison of a calculated mean against a known standard. One example of this is in testing whether the index of dispersion (measured as the variance:mean ratio of quadrat data such as that examined in box 4.5, p. 56) differs from the value of one that is expected when the distribution is random. Values significantly different from one occur when the distribution is even or

Box 5.2 One-sample t test for testing whether a distribution is significantly non-random using quadrat methods

[1] $SE = \sqrt{\frac{2}{n-1}}$

[2] $t = \frac{|\text{ index of dispersion} - 1|}{SE}$

[3] $\text{df} = n - 1$

Where: SE is the standard error of the difference between the index of dispersion and one;
n is the number of data points;
$|$ index of dispersion $- 1|$ is the absolute difference (that is, converting any negative values to positive) between the index of dispersion and one;
the index of dispersion is the ratio between the variance and the mean (see box 4.5);
df are the degrees of freedom.

From the data on the distribution of 25 solitary wasp nests in box 4.5 we found that the index of dispersion was greater than one, and therefore there was a tendency for wasp nests to show a clumped distribution. To establish whether the distribution is significantly clumped, we need to use a one sample t test to identify whether the index of dispersion is significantly different from one.

1. Calculate the standard error of the difference between the index of dispersion and one [1]	$SE = \sqrt{\frac{2}{n-1}} = \sqrt{\frac{2}{25-1}} = 0.2887$
2. Calculate the t statistic using the index of dispersion of 1.289 (see box 4.5) [2]	$t = \frac{\lvert\text{ index of dispersion} - 1\rvert}{SE} = \frac{\lvert 1.289 - 1\rvert}{0.2887} = 1.00$
3. Calculate the degrees of freedom [3]	For a one sample t test the degrees of freedom $= n - 1 = 25 - 1 = 24$
4. Look up the probability (P). See table A2 in Appendix II	For $t = 1.00$, $\text{df} = 24$, $P > 0.05$ (computer programs give an exact value of $P = 0.327$)
5. Do not reject the null hypothesis if the calculated probability is greater than or equal to the critical level OR reject the null hypothesis if the calculated probability is less than the critical level and examine the index of dispersion to determine whether the distribution is even or clumped	Here P is greater than the critical level. Therefore we do not reject the null hypothesis and we can say that: "The distribution of nest sites of the solitary wasp studied was not significantly different from a random distribution (index of dispersion = 1.289, $t = 1.00$, df = 24, $P > 0.05$)." Note that we have not demonstrated that the distribution is random; we have simply failed to show that it is significantly non-random. This subtle distinction is important because we are testing the null hypothesis of no difference from a random distribution. Yeo & Corbet (1995, NH 3) give more information about the ecology of solitary wasps.

Box 5.3 z test for testing whether a distribution is significantly non-random using nearest neighbour methods

[1] $SE_r = \frac{0.26136}{\sqrt{n\ d}}$

[2] $z = \frac{|r_d - r_e|}{SE_r}$

Where: SE_r is the standard error of the expected distance to the nearest neighbour;
$|r_d - r_e|$ is the absolute difference (that is, converting any negative values to positive) between the mean distance to the nearest neighbour (r_d) and the expected distance between nearest neighbours if the distribution is random (r_e);
n is the number of individuals in the study area;
d is the density of organisms (in the square of the units used for the nearest neighbour measurements).

From the data on the distribution of 20 lugworm burrows in box 4.6 we found that the index of aggregation was 0.7615, and therefore there was a tendency towards a clumped distribution. To establish whether there is a significantly clumped distribution, we need to use a z test to identify whether there is a significant difference between the measured and expected distances between nearest neighbours.

1. Calculate the standard error (using the density 0.000042 per square millimetre from box 4.6) [1]	$SE_r = \frac{0.26136}{\sqrt{n\ d}} = \frac{0.26136}{\sqrt{20 \times 0.000042}} = 9.02$
2. Calculate the test statistic (using the mean distance of 58.75 mm and expected distance of 77.1517 mm from box 4.6) [2]	$z = \frac{\lvert r_d - r_e \rvert}{SE_r} = \frac{\lvert 58.75 - 77.1517 \rvert}{9.018} = 2.04$
3. Look up the probability (P). See table A2 in Appendix II. Note that for the z test the degrees of freedom are always infinity (∞)	For $z = 2.04$, $P < 0.05$ (computer programs give an exact value of $P = 0.0413$)
4. Do not reject the null hypothesis if the calculated probability is greater than or equal to the critical level OR reject the null hypothesis if the calculated probability is less than the critical level and examine the index of aggregation to determine whether the distribution is clumped or even	Here P is lower than the critical level. Therefore we reject the null hypothesis and, since the index of aggregation is less than 1 (0.7615), we can say that: "Lugworm burrows showed a significantly clumped distribution (index of aggregation = 0.7615, $z = 2.04$, $P < 0.05$)." This may mean that there is a direct influence of the position of one lugworm burrow on another, with lugworms selecting sites near to each other for some reason. Alternatively, it could be an indirect effect if appropriate conditions for burrowing (for example, density of sand, availability of food) were patchy so that individual lugworms aggregated to exploit these. Hayward (1994, NH 21) gives further information about the distribution of lugworm burrows.

clumped. The test statistic is again t. Box 5.2 shows the calculation of t in relation to the example of the distribution of solitary wasps' nests.

A test closely related to the t test, called a z test, may also be used to examine whether a distribution is random, even or clumped, but using nearest neighbour data (pp. 56–58) rather than quadrat data. Box 5.3 shows the calculation of the test statistic (z) using data on the distribution of lugworm burrows. Since we found that the index of aggregation was less than one (box 4.6) we suspect that the burrows are clumped together. The z test allows us to compare the actual mean distance recorded with that expected if the distribution was random.

Mann–Whitney *U* test

This is a nonparametric equivalent of the t test suitable for ranked and non-normal data. As in many other nonparametric tests, the procedure, if carried out by hand, involves first ranking the data. The data from both samples together are placed in a single list in ascending order and each data point is allocated a rank based on its position. So the values 6, 3, 7, 4, 9 would be ordered as 3, 4, 6, 7, 9. The smallest value (3) is now allocated a rank of 1, the next smallest (4) is ranked 2, and so on. This process gets a little more complicated if we have several data points with the same value. Now data points share ranks as follows. If we have ten data points placed in order of 3, 4, 4, 6, 7, 7, 7, 7, 7, 9 then the smallest value (3) is allocated the rank of 1, but because the next two data points have the same value, they share the next two ranks (rank 2 and 3) giving each of them a rank of 2.5, calculated as (2+3)/2. The next data point (6) is allocated a rank of 4, the next five share the ranks of 5, 6, 7, 8, and 9 and are each ranked 7, calculated as (5+6+7+8+9)/5. Finally the highest value is ranked 10. The sums of these ranks in each sample, together with the numbers of data points, in each sample are used to generate two test statistics (U_1 and U_2). The smaller of these two statistics together with the number of data points in each sample (n_1 and n_2) is then used to generate a probability which is compared with the critical value of 0.05. If a significant difference between samples is obtained, it is necessary to examine the rank means of the samples (the sum of the ranked values in a sample divided by the number of data points in that sample) to identify where the difference lies.

One drawback of this test when done by hand in this way is that when there are many tied ranks (that is when a particular value occurs more than once), the test is conservative and may give non-significant results even when differences are present. In such situations, if the probability is only slightly larger than the critical value, it may be worth applying a correction factor. Zar (1999) gives details.

The worked example in box 5.4 compares the numbers of thrips that are found in two different types of

Box 5.4 Mann–Whitney *U* test

$$[1]\ U_1 = n_1 n_2 + \frac{n_1(n_1+1)}{2} - \Sigma R_1$$

$$[2]\ U_2 = n_1 n_2 + \frac{n_2(n_2+1)}{2} - \Sigma R_2$$

Where: n_1 and n_2 are the number of data points in samples 1 and 2 respectively;
ΣR_1 and ΣR_2 are the sums of the ranks in samples 1 and 2.

If the arithmetic has been correctly performed: $[3]\ U_1 + U_2 = n_1 n_2$

In an investigation of the efficiency of water traps for collecting thrips, ten traps containing water were compared with ten similar traps containing water to which a floral scent had been added. Because one trap contained a large number of thrips that were not accurately counted, but simply recorded as over 100 animals, a *t* test is inappropriate and a Mann–Whitney *U* test should be used.

1. Record the data.
   Note the data point with over 100 thrips has been allocated a value of 101

Numbers of thrips found in each of ten water traps (w):
25 42 18 16 7 33 57 84 92 22
Numbers of thrips found in each of ten water trap with floral scent (f):
55 44 36 29 67 72 83 101 61 64

2. Compile the samples into a single ordered list and rank the data using mean ranked values when tied values occur (see text for details)

Trap type	w	w	w	w	w	f	w	f	w	f	f	w	f	f	f	f	f	w	w	f
Number	7	16	18	22	25	29	33	36	42	44	55	57	61	64	67	72	83	84	92	101
Rank	1	2	3	4	5	6	7	8	9	10	11	12	13	14	15	16	17	18	19	20

3. Calculate the rank sums and rank means for each of the two samples

The rank sums are: $R_w = 1 + 2 + 3 + 4 + 5 + 7 + 9 + 12 + 18 + 19 = 80$
$R_f = 6 + 8 + 10 + 11 + 13 + 14 + 15 + 16 + 17 + 20 = 130$

The rank means are: for water traps = 80/10 = 8.0
for water traps plus floral scent = 130/10 = 13.0

Box 5.4 (continued)	
4. Calculate the U statistics and choose the smaller of the two values [1] and [2]	$U_w = n_w n_f + \frac{n_w(n_w+1)}{2} - \sum R_w = 10\times10 + \frac{10\times(10+1)}{2} - 80 = 75$ $U_f = n_w n_f + \frac{n_f(n_f+1)}{2} - \sum R_f = 10\times10 + \frac{10\times(10+1)}{2} - 130 = 25$
5. Check the calculations [3]	$U_w + U_f = n_w n_f$ $\quad 75 + 25 = 100$ and $10 \times 10 = 100$
6. Look up the probability (P) for the smaller U value. See table A3 in Appendix II	For $U = 25$ with $n_w = 10$ and $n_f = 10$, $P > 0.05$ (computer programs give an exact value of $P = 0.059$)
7. Do not reject the null hypothesis if the calculated probability is greater than or equal to the critical level OR reject the null hypothesis if the calculated probability is less than the critical level and examine the rank means to determine where the difference lies	Since P is greater than the critical level, we do not reject the null hypothesis and can say that: "There is no significant difference in the number of thrips caught in plain water and floral scented water traps ($U = 25$, $n_w = 10$, $n_f = 10$, $P > 0.05$)." There may indeed be no differences in the numbers of thrips in the two types of water trap, or the relatively small data set may have concealed a relatively small difference between the numbers. Further investigations may clarify this. Kirk (1996, NH 25) gives details of thrips and their ecology.

Box 5.5 Paired *t* test

[1] $\bar{d} = \frac{\Sigma d}{n}$ [2] $SE_d = \sqrt{\frac{\Sigma d^2 - \frac{(\Sigma d)^2}{n}}{n(n-1)}}$

[3] $t = \frac{|\bar{d}|}{SE_d}$ [4] $df = n - 1$

Where: $\bar{d}$ is the mean of the differences between pairs of data;
d is the difference between each pair of data points;
n is the number of pairs of data points;
SE_d is the standard error of the mean difference;
df are the degrees of freedom.

The numbers of seven spot ladybirds (*Coccinella 7-punctata*) on two different types of plants (stinging nettles and grasses) were compared. Ten sites were chosen, each of which contained a patch of nettles and a similar sized patch of grasses. The plants were sampled by observations lasting for 30 minutes in each patch and the numbers of seven spot ladybirds were recorded. Differences in the numbers of ladybirds occurring on nettles and grasses are examined using paired *t* tests because the data are matched by site (each site having two patches of plants – one of nettles and the other of grasses).

1. Record the data

Site:	A	B	C	D	E	F	G	H	I	J
Numbers on nettles:	98	87	66	55	102	82	71	28	66	55
Numbers on grasses:	75	79	67	43	107	54	70	24	57	52

2. Calculate the differences between pairs, then square each difference. Find the sum of the differences. Sum the differences between pairs and divide by the number of pairs (10) to find the mean difference [1]

Differences (d):	23	8	–1	12	–5	28	1	4	9	3
Squares of the differences (d^2):	529	64	1	144	25	784	1	16	81	9

$\Sigma d = 23 + 8 - 1 + 12 - 5 + 28 + 1 + 4 + 9 + 3 = 82$

$\Sigma d^2 = 529 + 64 + 1 + 144 + 25 + 784 + 1 + 16 + 81 + 9 = 1654$

$$\bar{d} = \frac{\Sigma d}{n} = \frac{82}{10} = 8.2$$

Box 5.5 (continued)	
3. Calculate the standard deviation of the mean difference [2]	$SE_d = \sqrt{\dfrac{\Sigma d^2 - \dfrac{(\Sigma d)^2}{n}}{n(n-1)}} = \sqrt{\dfrac{1654 - \dfrac{(82)^2}{10}}{10(10-1)}} = 3.30$
4. Calculate the t statistic [3]	$t = \dfrac{\lvert\bar{d}\rvert}{SE_d} = \dfrac{8.2}{3.025} = 2.48$
5. Calculate the degrees of freedom [4]	$df = n - 1 = 10 - 1 = 9$
6. Look up the probability (P). See table A2 in Appendix II	For $t = 2.48$ with df = 9, $P < 0.05$ (computer programs give an exact value of $P = 0.0348$)
7. Do not reject the null hypothesis if the calculated probability is greater than or equal to the critical level OR reject the null hypothesis if the calculated probability is less than the critical level and examine the sign of the mean difference to determine where the difference lies	Here P is less than the critical level. Therefore we reject the null hypothesis and, since subtracting the number on grasses from the number on nettles results in a positive mean difference (step 2) we can say that: "There were significantly more seven spot ladybirds in patches of nettles than there were in grass patches ($t = 2.48$, df = 9, $P < 0.05$)." These differences may be due to direct effects (e.g. contrasts between the structure of the two types of plants), or indirect effects (e.g. the numbers of prey animals such as aphids that frequent the different types of plants). Note that this test controls for the fact that the number of ladybirds varies between sites, regardless of plant species (e.g. site H has fewer ladybirds in either plant type than does any other site). Davis (1991, NH 1) and Majerus & Kearns (1989, NH 10) give further details about ladybirds on nettles.

traps. We might consider using a *t* test for these data (assuming that they are normally distributed). However, in this instance, one trap contains a huge number of thrips and is simply recorded as having 100+ thrips. This makes it impossible to calculate a mean, so a *t* test cannot be used. However a median can be calculated and the data are rankable, so a Mann–Whitney *U* test is appropriate. Table 4 (p. 68) illustrates how to consult a table of *U* values to find the probability for this example.

The Mann–Whitney *U* test may also be useful when counts of relatively rare events are being compared between two samples, such as the numbers of periwinkles found on seaweed fronds belonging to two different species. In such cases the data are often not normally distributed. This is because there are likely to be a large number of fronds with no periwinkles, rather fewer with one or two periwinkles, and even fewer with larger numbers of periwinkles (giving a distribution similar to that in fig. 95). To compare the numbers of periwinkles found on two different species of seaweed, then, with non-normal data, a *t* test is inappropriate but a Mann–Whitney *U* test can be used. In situations where a variable, such as degree of leaf damage caused by insect herbivory, is measured on an approximate scale (perhaps to the nearest 10%) then the data are not measurements (that is, interval or ratio data), but they can be ranked. Therefore to compare beetle damage on dock plants in exposed and sheltered situations (Salt & Whittaker, 1998, NH 26), we would employ a Mann–Whitney *U* test.

Paired *t* test

When data are suitable for *t* tests (that is, normally distributed measurements) but derive from project designs producing matched data (p. 5), then a variant called a paired *t* test should be used. For example, measurement of temperatures on the upper and lower surfaces of a series of leaves (Unwin & Corbet, 1991, NH 15) produced data paired by leaf. Here the differences examined are those between the matched pairs of data rather than between the sample means as in an ordinary *t* test. Hence, if any difference is shown to be significant, it is the direction of the mean difference between pairs of data that is used in the interpretation of the results. Box 5.5 uses data on seven spot ladybirds in adjacent patches of nettles and grasses to illustrate the calculations for this test.

A paired *t* test is more powerful (that is, more likely to identify significant differences) than an ordinary *t* test. For example, in box 5.5 some of the variation between sites is controlled by taking pairs of readings from each site. That is why it can be worthwhile to design an investigation in such a way as to produce paired data points (p. 5). Other investigations in which the paired *t* test is useful include, for example, a comparison of the numbers of insects (such as mayflies, Harker, 1989, NH 13) that are intolerant to organic pollution found above and below drains flowing into several urban streams. The data are paired by stream, each having two samples, one from above and one from below the outflow. Mason (1996) gives more information on pollution studies in freshwaters. Similarly, a paired *t* test is appropriate in examining differences in the numbers of organisms found on the exposed and sheltered sides of several individual trees (for example, oak spangle galls, Redfern & Askew, 1992, NH 17, or lichens, Richardson, 1992, NH 19) where each tree is used twice, once for an exposed sample and once for a sheltered one. Investigations into the animals found under logs or stones at the edges and centres of several woodlands (Wheater & Read, 1996, NH 22) provide a pair of samples (one from the centre and one from the edge) for each woodland.

Wilcoxon matched-pairs test

In the same way that the Mann–Whitney *U* test is a nonparametric equivalent of the *t* test, the Wilcoxon matched-pairs test is a nonparametric equivalent of the paired *t* test. Here, after finding the difference between each pair of scores, the differences are ranked and used to generate two test statistics (T_+ and T_-) by summing the ranks of those pairs with positive differences and separately summing those with negative differences. If a significant difference is found, the sizes of the sum of ranks of the positive and negative differences are examined to determine where the differences lie. Box 5.6 shows the calculation of the Wilcoxon matched-pairs test using data on foraging behaviour in ground beetles.

The Wilcoxon matched-pairs test would also be useful in, for example, a comparison of the animals found on a range of sandy shores just before and just after high tide. This would require a paired test, matching the data by shore. If many highly mobile animals were recorded, it might be difficult to count them accurately, and estimates might be needed. In this case, the data are not measurements, but if they are rankable the Wilcoxon matched-pairs test would be appropriate. Where there are many tied ranks this test is conservative. Thus if the probability is only slightly larger than the critical value and values of the differences between pairs are tied, it may be worth applying a correction factor. Zar (1999) gives details.

Analysis of variance

Where more than two samples are being compared, it might seem logical to test the differences between pairs of samples (for example when contrasting three sites, to compare sites A and B, then A and C and finally B and C). However, this is not appropriate because the statistical tests

Box 5.6 Wilcoxon matched-pairs test

[1] T_+ = the sum of the ranks of the positive differences between pairs of data points

[2] T_- = the sum of the ranks of the negative differences between pairs of data points

As a check of the correct calculation of T_+ and T_-:

[3] $T_+ = \frac{n(n+1)}{2} - T_-$ and [4] $T_- = \frac{n(n+1)}{2} - T_+$

Where: n = the number of pairs of data points.

In ground beetles, foraging behaviour may depend on whether the animals are aware of potential prey in the locality or not. This was studied in *Pterostichus madidus* (a commonly found ground beetle). The rate at which animals changed direction (measured as the number of turns per metre moved) was examined in animals that had been starved for seven days, and in the same animals that had been briefly exposed to a prey item (a blowfly larva for 30 seconds). The data are matched by animal and the differences appear not to conform to a normal distribution (see Chapter 4). Thus comparisons between the animals when starved and after they have been recently exposed to prey may be made using a Wilcoxon matched-pairs test.

Step		1	2	3	4	5	6	7	8	9	10
	Beetle number	1	2	3	4	5	6	7	8	9	10
1. Record the data	Starved (turns per m moved)	30.2	26.1	24.6	24.5	27.4	30.8	25.4	24.7	25.1	27.2
	Exposed to prey (turns per m moved)	30.2	31.7	30.3	30.8	30.1	30.2	32.2	30.6	32.1	31.6
2. Find the differences between matched pairs. Subtract the scores for one sample from the corresponding score in the other	Differences between matched pairs	0	–5.6	–5.7	–6.3	–2.7	+0.6	–6.8	–5.9	–7.0	–4.4
3. Rank the differences ignoring the sign. Zero differences are ignored, and mean rank values are used if more than one of a particular value occurs. Label with the sign of the difference	Ranks		4	5	7	2	1	8	6	9	3
	Sign of the difference		–	–	–	–	+	–	–	–	–

Box 5.6 (continued)	
4. Calculate the number of data pairs (not including any zero differences)	Omitting zero differences there are now nine data pairs (n = 9), of which one is a positive difference and eight are negative differences.
5. Calculate the test statistics by summing the ranks of the negative and positive differences separately [1] and [2] Check the calculations using formulae [3] and [4]	$T_+ = 1$ and $T_- = 4 + 5 + 7 + 2 + 8 + 6 + 9 + 3 = 44$ $T_+ = \frac{n(n+1)}{2} - T_- = \frac{9(9+1)}{2} - 44 = 1$ $T_- = \frac{n(n+1)}{2} - T_+ = \frac{9(9+1)}{2} - 1 = 44$
6. Select the smaller of the two T values	$T_+ = 1$
7. Look up the probability (P). See table A4 in Appendix II	For $T_+ = 1$ with $n = 9$, $P < 0.05$ (computer programs give an exact value of $P = 0.0109$)
8. Do not reject the null hypothesis if the calculated probability is greater than or equal to the critical level OR reject the null hypothesis if the calculated probability is less than the critical level and examine which rank of the summed difference (T_+ or T_-) is the larger to determine where the difference lies	Here P is less than the critical level. Therefore we reject the null hypothesis. Subtracting the number of turns made by animals exposed to prey from the number of turns made by starved animals more often results in a negative value and so the sum of the ranks of the negative differences is the higher. That is, animals exposed to prey have a higher turn rate, and we can say that: "Starved beetles that have been briefly exposed to prey exhibit a greater turning rate than do starved individuals before they have been exposed to prey ($T_+ = 1$, $n = 9$, $P < 0.05$)." This may indicate that animals that have been exposed to prey are concentrating their foraging behaviour on a smaller area (perhaps to exploit possible patches of prey items) than are those animals that were unexposed to prey. Forsythe (2000, NH 8) gives more information about ground beetle ecology and feeding behaviour.

used to compare two samples become unreliable when used in this way. Although beyond the scope of this book, there is a suite of tests available for comparing more than two samples, and more complex designs, such as those that look at more than one factor at a time. For example, if we wished to look at the numbers of insects caught in colour traps of two different sizes and three different colours, we could compare the sizes and colours separately and as a combination. There are parametric versions (analysis of variance, known as ANOVA) and nonparametric versions (analysis of variance using ranks) for both matched and unmatched data. Wheater & Cook (2000) describe these and explain how to use them.

Testing for relationships between variables

There are basically two mathematically similar statistical procedures for testing for relationships: correlation and regression. Correlation analysis is used to identify relationships between variables, whereas regression analysis produces an equation that describes a supposedly causal relationship and may be used to predict values of the dependant variable from new values of the independent variable (p. 7). This key identifies tests for relationships between only two variables; when testing for relationships between more than two variables see texts such as Zar (1999) or Sokal & Rohlf (1995). It is important to check the shape of any relationship by using a scatterplot of one variable against the other (fig. 107). None of the techniques described below are appropriate for relationships shaped like $\subset$ or $\supset$ or $\cap$ or $\cup$. The two parametric methods (Pearson's product moment correlation and regression analysis) are appropriate only if there is an approximately straight line relationship.

Key B Statistical tests for relationships

1 Both variables are measurements (that is, interval or ratio data) — 2

– One or both variables are measured on a ranked scale — Spearman's rank correlation (p. 85)

2 Both variables approximate to a normal distribution — 3

– One or both variables do not approximate to a normal distribution — Spearman's rank correlation (p. 85)

3 Simple correlation required — Pearson's product moment correlation (p. 85)

– Predictive model required — Regression analysis

See p. 87 for a brief description and Wheater & Cook (2000) for more information

Pearson's product moment correlation

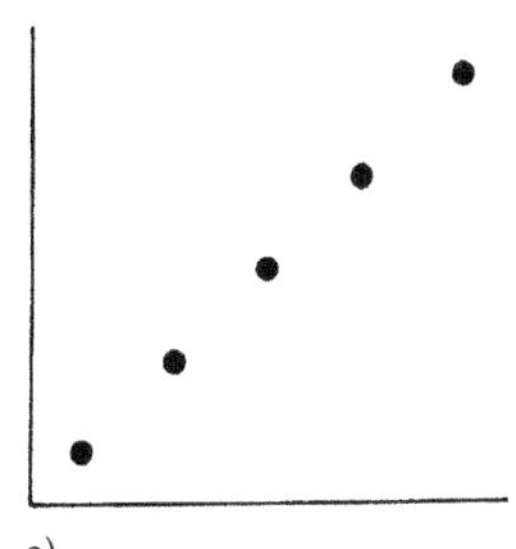
a)

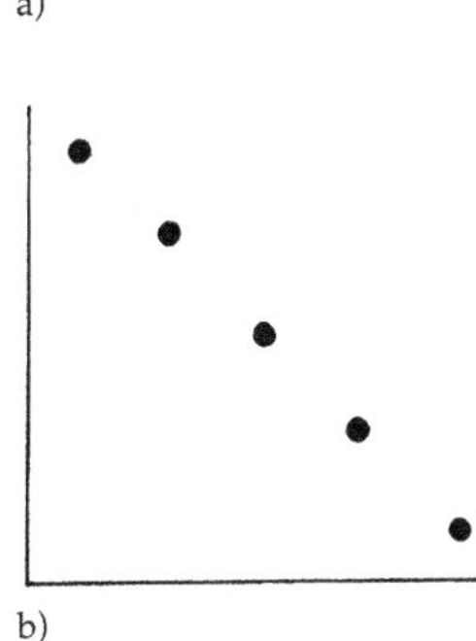
b)

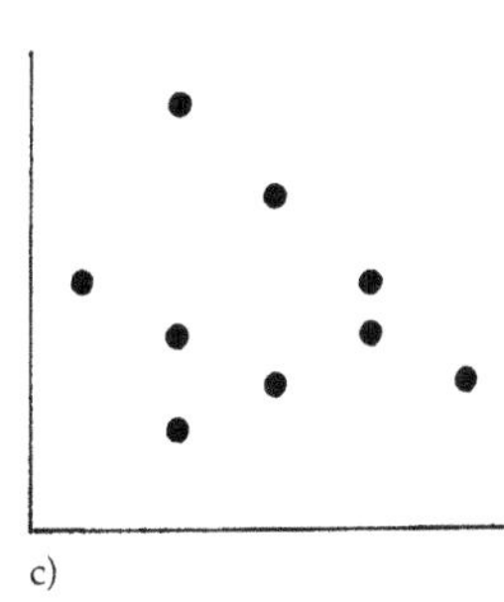
c)

Fig. 103. Scatterplot showing range of possible correlation coefficients:
a) perfect positive correlation ($r = +1$)
b) perfect negative correlation ($r = -1$)
c) no correlation ($r \sim 0$)

To see whether there is a simple relationship between variables that are both measurements conforming to a normal distribution, we need a parametric test. Pearson's product moment correlation is the appropriate test here. A coefficient (r) is calculated that can vary between –1 and +1 (fig. 103). A value close to +1 indicates a positive relationship (that is, as one variable increases so does the other variable), while a value near to –1 indicates a negative relationship (as one variable increases, the other decreases). Values around 0 indicate that there is no particular systematic variation of one variable with the other. To find out how close to +1 (or –1) a correlation coefficient has to be in order to claim a significant positive (or negative) relationship, we use the correlation coefficient and the sample size (degrees of freedom) to generate the probability of obtaining these results if the null hypothesis is correct. An example showing how Pearson's product moment correlation is calculated is given in box 5.7, using data about the relationship between the air temperature and the time relative to sunset at which mayflies stopped swarming.

Examples of some other situations in which Pearson's product moment correlation might be useful include dragonfly territorial behaviour. Many male dragonflies defend their territories against non-residents. If a male is removed from his territory and held captive, another dragonfly may move in and defend the territory against other males, including the original territory holder (Miller, 1995, NH 7). To investigate this behaviour we might examine the relationship between the time the original resident was held captive and the time he spends trying to regain his lost territory. The appropriate test would be a Pearson's product moment correlation.

Another interesting relationship to study is that between the size of a habitat and the number of animal species. Such a relationship may not be a strictly linear one. For example, in many island situations (including habitat islands such as isolated trees or clumps of plants) a tenfold increase in the island size approximately doubles the number of species (Leather & Bland, 1999, NH 27). However, the number of individual animals may also be related to the size of the habitat and this could be tested using Pearson's product moment correlation. For example, the possible relationship between the numbers of oak spangle galls and the size of the leaf could be examined (Redfern & Askew, 1992, NH 17) if gall counts and leaf measurements are normally distributed.

Spearman's rank correlation

This test is a nonparametric equivalent of the Pearson's product moment correlation for use when one or both of the variables are measured on a ranked scale, or are not normally distributed. Like the parametric statistic (r), the

Box 5.7 Pearson's product moment correlation coefficient

[1] $$r = \frac{n\sum xy - \sum x \sum y}{\sqrt{[n\sum x^2 - (\sum x)^2][n\sum y^2 - (\sum y)^2]}}$$

[2] $\mathrm{df} = n - 2$

Where: x and y represent pairs of data;
n is the number of data pairs;
Σx and Σy are the sums of x and y respectively;
Σxy is the sum of the products of x and y (i.e. each value of x multiplied by its associated value of y and then all summed);
df are the degrees of freedom.

The swarming behaviour of a species of mayfly was investigated to determine whether there was a relationship between the air temperature and the time (relative to sunset) at which swarming ceased. Temperature was taken 1.5 m above ground level, and the time that swarming ceased was calculated as the number of minutes after sunset.

1. Record the data

Temperature (C) (x)	9.3	9.9	11.2	12.3	13.3	16.2	16.6	16.7	18.1	21.2	22.3
Time (min) (y)	34	58	74	91	114	79	96	135	102	114	121

2. Calculate the values of xy, x^2, and y^2

xy	316.2	574.2	828.8	1119.3	1516.2	1279.8	1593.6	2254.5	1846.2	2416.8	2698.3
x^2	86.49	98.01	125.44	151.29	176.89	262.44	275.56	278.89	327.61	449.44	497.29
y^2	1156	3364	5476	8281	12996	6241	9216	18225	10404	12996	14641

3. Calculate the sums of x, y, xy, x^2, and y^2

$\Sigma x = 167.1$ $\Sigma y = 1018$ $\Sigma xy = 16443.9$
$\Sigma x^2 = 2729.35$ $\Sigma y^2 = 102996$

4. Calculate the r statistic [1]

$$r = \frac{n\sum xy - \sum x \sum y}{\sqrt{[n\sum x^2 - (\sum x)^2][n\sum y^2 - (\sum y)^2]}}$$

$$= \frac{(11 \times 16443.9) - (167.1 \times 1018)}{\sqrt{[(11 \times 2729.35) - 167.1^2][(11 \times 102996) - 1018^2]}}$$

$$= \frac{10775.1}{\sqrt{2100.44 \times 96632}} = \frac{10775.1}{14246.744} = 0.756$$

5. Calculate the degrees of freedom [2]

Degrees of freedom = number of data pairs − 2 = 11 − 2 = 9

6. Look up the probability (P). See table A5 in Appendix II

For $r = 0.756$ and df = 9, $P < 0.01$ (computer programs give an exact value of $P = 0.007$)

7. Do not reject the null hypothesis if the calculated probability is greater than or equal to the critical level
OR reject the null hypothesis if the calculated probability is less than the critical level and examine the sign of the r value to determine the direction of the relationship

Here P is less than the critical level. Therefore we reject the null hypothesis and, since the sign of the r value is positive, we can say that:

"There was a significant positive relationship between the air temperature and the time of cessation of swarming relative to sunset ($r = 0.756$, df = 9, $P < 0.01$)."

This means that the higher the air temperature, the later after sunset the mayflies stopped swarming. This may be a direct effect of air temperature on the behaviour of the mayflies. Alternatively there could be an indirect effect if, for example, air temperature is related to light level, which in turn could be a determinant of swarming behaviour. Harker (1989, NH 13) gives details of the impacts of weather on swarming in mayflies, and Unwin & Corbet (1991, NH 15) give more information about microclimate and insects.

test statistic (the correlation coefficient r_S) varies between –1 and +1. When calculating by hand, each variable is first ranked separately and then the absolute difference between each pair of data is found. The absolute difference involves subtracting one value from the other and ignoring any minus signs. Note that if there is more than one data point within a variable with the same value, then they share ranks (p. 75). A worked example is given in box 5.8 using data on the relationship between the percentage attack of thistle leaves by weevil larvae, and the size of the thistles.

Examples of other situations in which Spearman's rank correlation is appropriate include studies of mosquito larvae escape responses. Mosquito larvae come to the water surface to breathe. These larvae submerge quickly when disturbed, and then rise passively to the surface once again. Experiments might examine the relationship between the degree of disturbance and the numbers of larvae that submerge. If the disturbance is not quantifiable except by using a scoring system (such as no disturbance, slight, moderate, major, or extreme disturbance), Spearman's rank correlation is suitable. Guthrie (1989, NH 12) considers animals inhabiting the surface film, and Snow (1990, NH 14) gives information on mosquitoes.

Correlation analysis can also be used to examine whether the proportion of sheep with blowfly strike is related to the size of flock (Erzinçlioğlu, 1996, NH 23). If the proportion of sheep struck is an estimate, or if it is skewed towards either 0 or 1 (that is 0 or 100%), then a nonparametric test such as Spearman's rank correlation is required (see discussion of normal distributions p. 44). Similarly, if the sizes of flocks are only known approximately (up to 20 sheep, 20–50 sheep, and so on), Spearman's rank correlation is the appropriate test. Where there are numerous ties this test can give misleading results. In such cases, where the probability is only slightly lower than the critical value (that is, the relationship is only just significant) the true value may be higher (and the relationship therefore not significant). In these situations it may be worth applying a correction factor. Zar (1999) gives details.

Regression analysis

Correlation tells us whether there is a relationship between two variables. Regression does this and more: it allows us to predict the value of one variable from knowledge of the other. For example, we might wish to know the heights of trees in a dense woodland. It may be difficult to measure the heights using surveying techniques (fig. 5.5 on p. 145 in Wheater, 1999) if we cannot sight the tops of the trees. However, if there are a number of felled trees around, we can see whether there is a relationship between the height and girth of trees lying on the ground, and, if there is, use this to produce a model that allows us to estimate the height of standing trees from the girth. In a more

Box 5.8 Spearman's rank correlation coefficient

[1] $r_s = 1 - \frac{6\sum d^2}{(n^3 - n)}$

Where: d is the difference between the ranks within each pair of data points;
n is the number of data pairs.

As part of a study of the potential biological control of a thistle species using a weevil, thistles of different sizes were examined to identify the estimated percentage attack to the nearest 5% on the whole plant by weevil larvae. These data can be examined to ascertain whether there is a relationship between the size of plant and the percentage attack. Since the attack data are approximations rather than accurate measurements, Spearman's rank correlation coefficient is an appropriate method to use.

1. Record the data

Thistle size (m)	0.2	0.27	0.51	0.63	0.76	0.94
Percentage larval attack	10	25	45	40	50	55

2. Calculate the ranks of the data (that is, score each row of data from the lowest (rank 1) to the highest)

Rank of thistle size	1	2	3	4	5	6
Rank of larval attack	1	2	4	3	5	6

3. Calculate the differences between ranks

Differences between ranks	0	0	-1	1	0	0

4. Calculate the r_s statistic [1]

$$r_s = 1 - \frac{6\sum d^2}{(n^3 - n)} = 1 - \frac{6 \times (0^2 + 0^2 + (-1^2) + 1^2 + 0^2 + 0^2)}{(6^3 - 6)}$$

$$= 1 - \frac{12}{210} = 0.943$$

5. Look up the probability (P). See table A6 in Appendix II

For $r_s = 0.943$ and $n = 6$, $P < 0.05$
(computer programs give an exact value of $P = 0.035$)

6. Do not reject the null hypothesis if the calculated probability is greater than or equal to the critical level

   OR reject the null hypothesis if the calculated probability is less than the critical level and examine the sign of the r_s value to determine the direction of the relationship

Here P is less than the critical level. Therefore we reject the null hypothesis and, since the r_s value is positive, we can say that:

"There is a significant positive relationship between the size of thistle plants and the percentage attack by weevil larvae ($r_s = 0.943$, $n = 6$, $P < 0.05$)."

This result may be due to adult weevils selecting larger plants in which to lay their eggs, or may be a result of other indirect effects: for example, if larger plants grow in habitats where weevils are foraging for egg-laying sites. Redfern (1995, NH 4) gives further details of the potential for biological control of thistles.

experimental situation, we might wish to predict how many insect pests would be found on a crop treated with certain doses of pesticide. Instead of testing all possible doses, we could set up experiments that would look at, say, ten different doses and record the numbers of pests after spraying with each dose. If there is a straight line relationship between dose and pest count, we can use this information to predict the likely effect of any dose we might use in the future. Wheater & Cook (2000) explain how to carry out this type of analysis.

Testing for associations between frequency distributions

These tests are used to test the frequencies of several categories (that is, the frequency distribution of a categorical variable), either to see whether there is an association with another categorical variable, or for comparison with a known or theoretical expected distribution.

This key identifies tests that can be used when there are only one or two categorical frequency distributions. Texts such as Zar (1999) or Sokal & Rohlf (1995) show how to test for associations between more than two frequency distributions.

Key C Statistical tests for analysing frequencies

1 Testing two observed distributions against each other
Chi square test for association (below)

– Testing one observed distribution against a theoretical expected distribution
Chi square test for goodness of fit (p. 91)

Chi square test for association

The first step of this test takes the data from two categorical variables and puts them into cross-tabulated categories so that every combination of the categories is represented. If we had collected grasshoppers of different colours (two categories: green and brown) from different habitat types (three categories: lush, intermediate and dry), we could record how many green grasshoppers were found in lush habitats, how many in intermediate and how many in dry habitats, and then do the same for brown grasshoppers. This would make six (2 x 3 = 6) combinations. Using a table of these observed numbers, we can calculate the values we would expect if there is no association between grasshopper colour and habitat type (that is, if the null hypothesis is true), and compare these with the observed counts. Box 5.9 shows the calculation of a chi square test for association using data on grasshopper colour and habitat type. This test should not be used if the mean expected value (the total number of

Box 5.9 Chi square (χ^2) test for association

[1] $Expected\ value = \frac{Row\ total \times Column\ total}{Grand\ total}$

[2] $Cell\ \chi^2 = \frac{(Observed - Expected)^2}{Expected}$

[3] $Overall\ \chi^2$ = Sum of the cell χ^2 values

Where: the observed value is that obtained during the study;
the expected value is that predicted by a random distribution of items across the categories.

In a study of the colour dimorphism in grasshoppers, animals were collected from lush, intermediate and dry habitats and their colours classified as green or brown. The data can be examined using chi square tests to identify whether there is an association between grasshopper colour and habitat type.

1. Place the observed values in a contingency table

Colour of grasshopper	Numbers found in various habitat types			
	Lush	Intermediate	Dry	**Total**
Green	31	11	1	43
Brown	8	12	13	33
Total	39	23	14	76

2. Calculate the expected values as:

[1] $\frac{Row\ total \times Column\ total}{Grand\ total}$

e.g. for green on lush we have:
$(39 \times 43) / 76 = 22.07$

Colour of grasshopper	Numbers found in various habitat types			
	Lush	Intermediate	Dry	**Total**
Green	22.07	13.01	7.92	43
Brown	16.93	9.99	6.07	33
Total	39	23	14	76

3. Calculate the χ^2 statistic for each cell of the table:

[2] $\frac{(Observed - Expected)^2}{Expected}$

e.g. for green on lush we have:
$(31 - 22.07)^2 / 22.07 = 3.613$

Colour of grasshopper	Numbers found in various habitat types		
	Lush	Intermediate	Dry
Green	3.613	0.311	6.046
Brown	4.710	0.404	7.912

4. Calculate the overall χ^2 statistic as the sum of the cell χ^2 values [3]

χ^2 = the sum of the cell χ^2 values
$= 3.613 + 0.311 + 6.046 + 4.710 + 0.404 + 7.912 = 22.996$

5. Calculate the degrees of freedom

Degrees of freedom (df)
= (number of rows – 1) × (number of columns – 1)
$= (2 - 1) \times (3 - 1) = 2$

6. Look up the probability (P). See table A7 in Appendix II

For $\chi^2 = 23.00$ with 2 degrees of freedom, $P < 0.01$
(computer programs may show this as $P < 0.0001$)

7. Do not reject the null hypothesis if the calculated probability is greater than or equal to the critical level

OR reject the null hypothesis if the calculated probability is less than the critical level and examine the cell χ^2 values to determine where the associations lie

Here P is less than the critical level. Therefore we reject the null hypothesis and, since examining those cells with the large chi square values and comparing their observed and expected values, we can see that there are more brown grasshoppers on dry habitats than expected and more green grasshoppers on lush habitats than expected, we can say that:

"There is a significant association between grasshopper colour and habitat colour with green grasshoppers being more likely to occur on lush habitats and brown grasshoppers being likely to occur on dry habitats ($\chi^2 = 23.00$, df = 2, $P < 0.01$)."

The reasons for such findings may be habitat selection by grasshoppers or due to animals with contrasting (rather than cryptic) coloration being taken most frequently by visual predators such as birds, leaving behind animals that are harder to see. Brown (1990, NH 2) gives more information about crypsis in grasshoppers.

items divided by the number of combinations: box 5.9) is below six. If this does occur then the number of categories should be reduced by combining some together. Wheater & Cook (2000) give further details.

When both variables have only two categories, for example if we were investigating green and brown grasshoppers on lush and dry habitats only, this procedure gives a value of the test statistic (χ^2) that is slightly too high. We are therefore more likely to reject the null hypothesis (and claim a significant association) when in fact there is no such significant association. In this situation, when the calculated probability is only slightly smaller than the critical value ($P = 0.05$), it may be worth using a correction factor to take into account such inaccuracies. Wheater & Cook (2000) give further details.

Other examples where the chi square test for associations of frequency distributions would be useful include a comparison of the frequency distributions of the melanic (dark) and the non-melanic (normal red) forms of two-spot ladybirds in industrial cities and rural habitats (Majerus & Kearns, 1989, NH 10). A two-by-two table identifying the distributions of melanic and non-melanic forms across the two habitat types (rural and urban) could be analysed using a chi square test for association. Similarly, urban and rural sites could be compared with respect to the presence or absence of lichens intolerant to sulphur dioxide pollution (Richardson, 1992, NH 19). A study of male:male conflicts in thrips could include analysis of the relative size of challenger and defender (challenger larger or defender larger) against outcome (challenger wins or defender wins) (Kirk, 1996, NH 25).

Chi square test for goodness of fit

We might wish to compare the distribution of data over several categories with an expected theoretical distribution. Although the term 'theoretical distribution' sounds quite grand, it could simply be the expectation of an equal sex ratio. Here we use a goodness of fit test which works in a similar way to the chi square test for association. The major difference is that the goodness of fit test already has a series of expected values for comparison with the observed values from the study. Box 5.10 illustrates this test using data on sex ratios of spiders caught in traps, compared with the expected equal sex ratio.This test should not be used if the mean expected value (the total number of items divided by the number of combinations) is below two (or five if there are only two categories). If this does occur then the number of categories should be reduced by combining some together. Wheater & Cook (2000) give further details. Where there are only two categories, this procedure gives a value of the test statistic (χ^2) that is slightly too high. We are therefore more likely to reject the null hypothesis, and claim a significant departure from the expected distribution when in fact there is none. In this situation, if the calculated probability is only slightly smaller than the critical value ($P = 0.05$), it may be worth using a correction factor to take into account such inaccuracies. Wheater & Cook (2000) give further details. In our example in box 5.10, although there are two categories, the probability is greater than 0.05 so we do not need to use a correction factor.

The chi square test for goodness of fit of frequency distributions is also useful in, for example, comparing the distribution of greenflies on different parts of a leaf (such as along the edge, on the veins and in the area between the veins) with the expectation that the animals would be evenly distributed according to the relative surface area of the plant part concerned (Rotheray, 1989, NH 11). Plants and animals often colonise surfaces in a non-random fashion. Goodness of fit tests could be used to compare the distribution of barnacles on exposed and sheltered sides of rock faces with an expected distribution, or null hypothesis, of equal numbers on each type of face. Similarly, the frequencies of lichens found on north, east, south and west facing sides of tree trunks or gravestones could be compared with an expected 1:1:1:1 ratio (Richardson, 1992, NH 19).

Box 5.10 Chi square (χ^2) test for goodness of fit

[1] Cell $\chi^2 = \dfrac{(Observed - Expected)^2}{Expected}$

[2] Overall χ^2 = Sum of the cell χ^2 values

Where: the observed value is that obtained during the study; the expected value is calculated from a theoretical distribution.

Pitfall traps are a useful method of sampling surface active invertebrates (see Chapter 2). However, in a study of spiders in agricultural field margins, there was some concern that the captures of surface active wolf spiders were not representative of the population at large, because more male spiders (121) appeared to be captured compared with female spiders (98). A goodness of fit test enabled the data to be examined to see whether the frequency of male and female spiders differed significantly from that expected (1:1 ratio).

Step		Males	Females	**Total**
1. Record the observed counts		121	98	219
2. Calculate the expected values	Since we anticipate an equal sex ratio in nature, we expect equal numbers of males and females, so from 219 spiders we would expect 219/2 = 109.5 of each sex	109.5	109.5	219
3. Calculate the cell χ^2 statistic for each cell [1]	$\chi^2 = \dfrac{(Observed - Expected)^2}{Expected}$	1.208	1.208	
4. Calculate the overall χ^2 statistic as the sum of the cell χ^2 values [2]	χ^2 = the sum of the cell χ^2 values = 1.208 + 1.208 = 2.416			
5. Calculate the degrees of freedom	The degrees of freedom (df) = the number of cells – 1 = 2 – 1 = 1			
6. Look up the probability (*P*). See table A7 in Appendix II	For χ^2 = 2.42 with 1 degree of freedom, $P > 0.05$ (computer programs give exact values of 0.120)			
7. Do not reject the null hypothesis if the calculated probability is greater than or equal to the critical level OR reject the null hypothesis if the calculated probability is less than the critical level and examine the cell X^2 values to determine where the associations lie	Here *P* is greater than the critical level. Therefore we do not reject the null hypothesis and can say that: "The relative numbers of male and female wolf spiders caught using pitfall traps in agricultural field margins did not significantly differ from an expected 1:1 ratio (X^2 = 2.42, df = 1, $P > 0.05$)." For more details about pitfall sampling see Chapter 2 or texts such as Wheater & Read (1996, NH 22).			

6 Presenting your results

Creating tables and graphs

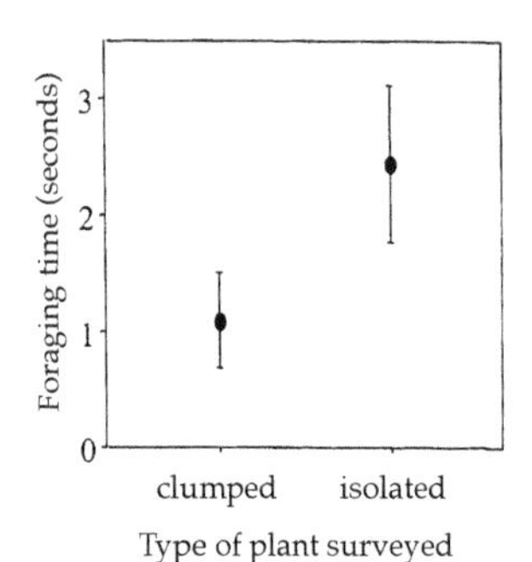

Fig. 104. Point chart of means of bumblebee foraging time on plants found in clumps ($n = 10$) compared with isolated plants ($n = 9$). Bars indicate standard errors.

Tables and graphs help us to visualise our results and to communicate our findings to others. First, we have to decide whether information is best given as a table, or as a figure, or simply described in the text. Use a table when there are many numerical data and where the exact numbers are important. Use a figure where graphical presentation reveals patterns that are not obvious from a table. In tables, 0 should indicate a reading of zero, not a missing data point. Tables can be used to present the results of repetitive statistical tests, as well as descriptive statistics (see table 5). Remember to report the test statistic, the number of data values (n or the degrees of freedom) and the probability, together with any relevant summary statistics such as means and standard errors.

Graphs, diagrams, maps and drawings are all described as 'figures'. Graphs are used for data that show trends or patterns, and for results you wish to highlight.

Table 5. *Mean numbers of insects found in one-litre samples taken from ten clean ponds and ten polluted ponds (with high levels of nitrates and phosphates). Numbers in brackets are standard deviations.*

	Polluted ($n = 10$)	**Clean ($n = 10$)**	***t* value (df = 18)**	***P***
Odonata	12.2 (2.13)	16.3 (1.35)	5.14	<0.0001
Plecoptera	1.3 (0.61)	4.6 (1.21)	7.70	<0.0001
Ephemeroptera	1.6 (0.83)	7.6 (2.01)	8.73	<0.0001
Trichoptera	3.1 (0.96)	5.7 (0.65)	7.09	<0.0001
Hemiptera	4.6 (1.63)	4.9 (2.10)	0.36	0.725
Diptera	35.6 (8.51)	16.4 (6.12)	5.79	<0.0001
Neuroptera	2.1 (0.81)	3.4 (1.25)	2.76	0.013
Coleoptera	6.4 (1.65)	7.6 (2.94)	1.13	0.275

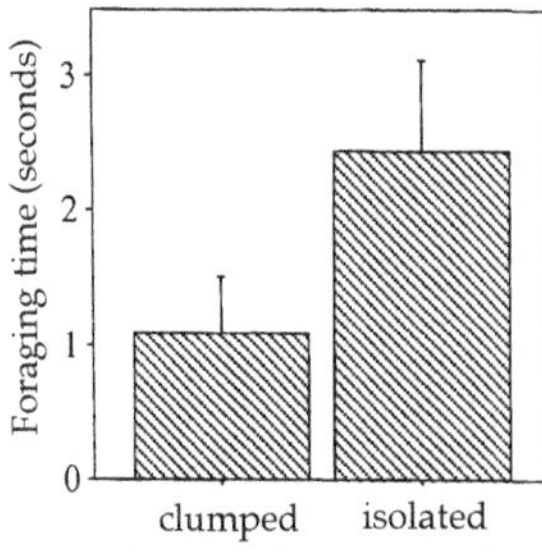

Fig. 105. Bar chart of means of bumblebee foraging time on plants found in clumps ($n = 10$) compared with isolated plants ($n = 9$). Bars indicate standard errors.

There are several types of graph that are used to display different types of data. Each of them must be accompanied by a descriptive text called the legend.

Histograms show frequency distributions of measured values, such as counts of eggs laid on brassica crops by hoverflies (box 4.1).

Point or bar charts (usually means with standard error bars or 95% confidence limits) are used to compare counts or measurements of two or more nominal variables. For example, in a study of the time budgets of bumblebees, the handling time per flower on plants growing in clumps might be compared with that on isolated plants (box 5.1 and figs. 104 and 105). Here each mean value is plotted as either the point or the top of the bar, and the vertical lines are used to represent the standard errors or 95% confidence limits. The figure legend must state which is used. Note the difference

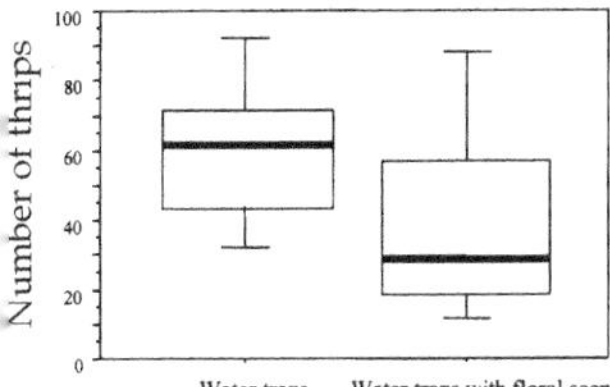

Fig. 106. Box and whisker plot of the numbers of thrips caught in water traps ($n = 10$) and water traps with a floral scent added ($n = 10$).

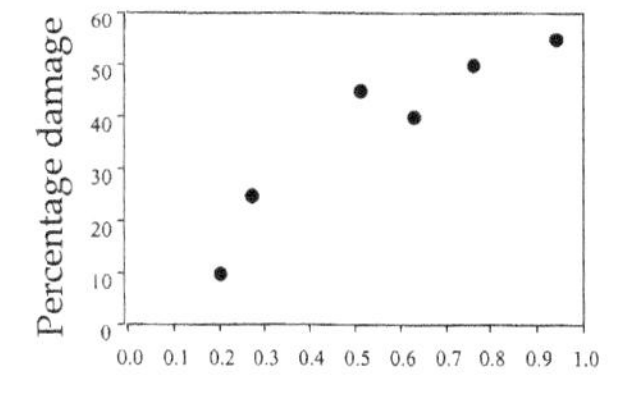

Fig. 107. Scatterplot of percentage damage caused by weevil larvae to thistles of various heights.

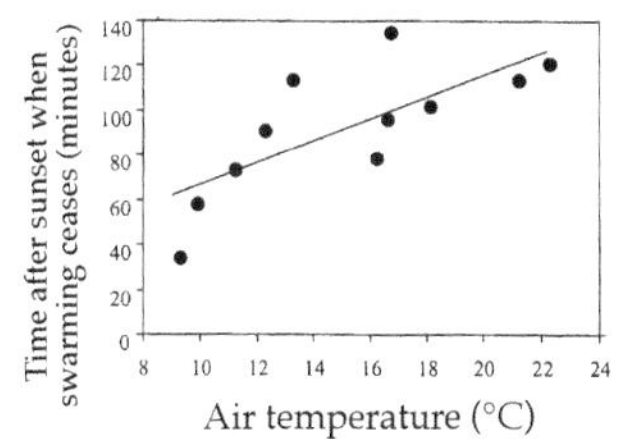

Fig. 108. Scatterplot showing regression line of time after sunset when swarming in mayflies ceases in relation to the air temperature.

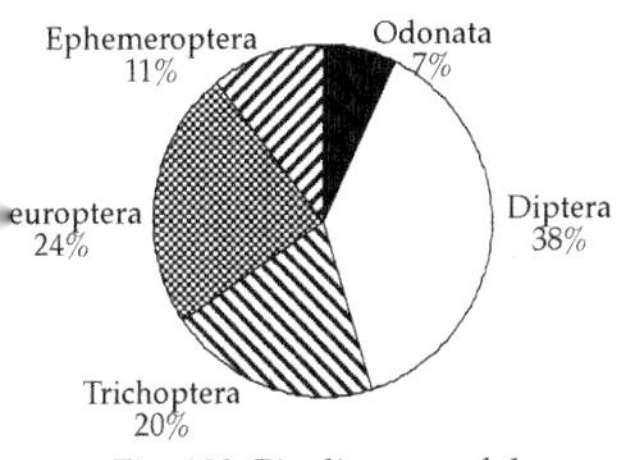

Fig. 109. Pie diagram of the proportion of insect orders found in a pond ($n = 185$).

between bar charts and histograms: bar charts have spaces between the bars, whereas histograms do not (box 4.1).

'Box and whisker' plots are used to compare median values (and interquartile ranges) of ranked scores from two or more samples. For example, in examining the ranked numbers of thrips caught in two different types of trap (box 5.4 and fig. 106), we used medians, rather than means, because one of the values was an estimate, rather than an exact measurement. The median is shown as a bold line, while the boxes indicate the interquartile range and the whiskers the full range of the data. Some computer programs extend the whiskers to cover a smaller range of data (for example from the 10% to the 90% points) and use asterisks to indicate outlier points. A box and whisker plot is a good way of illustrating the central tendency and variation within skewed data.

Scatterplots are used to show relationships between two measured or ranked variables, such as the percentage damage caused by weevil larvae to plants of different heights (box 5.8 and fig. 107). Calculated regression lines can be added to scatterplots when predictions are required, for example when predicting the time after sunset that mayflies will stop swarming, given a particular air temperature (see box 5.7 and fig. 108). Many computer graphics packages will calculate such a line. Wheater & Cook (2000) explain how to calculate regression lines by hand.

Pie charts show the proportional contributions of different categories to the whole, such as the proportions of different animal groups that make up insect fauna of the bottom of a pond (fig. 109). The segment labels can be the actual counts, or the proportions (or percentages), or both. Alternatively, if comparing two or more sites, the frequencies or percentages of each category of a nominal variable can be displayed in stacked or clustered bar charts (figs. 110 and 111).

Once data have been entered on a spreadsheet in a computer program such as Excel, Minitab or SPSS, graphs can quickly be produced. But keep your wits about you, and have a clear idea (or perhaps a sketch) of the graph you want before you ask a computer to produce it. Otherwise it is all too easy to generate nonsensical graphs. An easy to use Excel workbook is available from the authors (address p. 100) that illustrates the graphs produced using the data from the worked examples in this book and can be used to display your own data.

Every figure and table must have a legend that gives a full explanation of the contents, so that it can be understood by someone who has not read the text. Every figure and table must be numbered, and each must be cited in the text thus: (fig. 1). The figures and tables must be numbered, and appear, in the sequence in which they are cited in the text. By convention, we never cite fig. 2 before fig. 1. If that happens the figures or tables should be renumbered and reordered. Text should stand alone and refer to tables

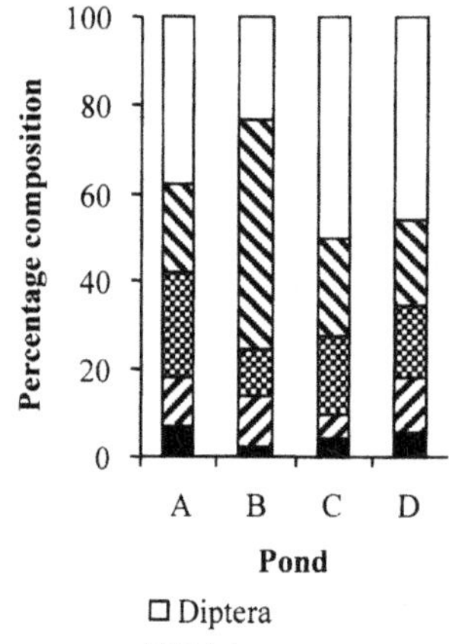

Fig. 110. Stacked bar graph of the proportion of insect orders found in four ponds.

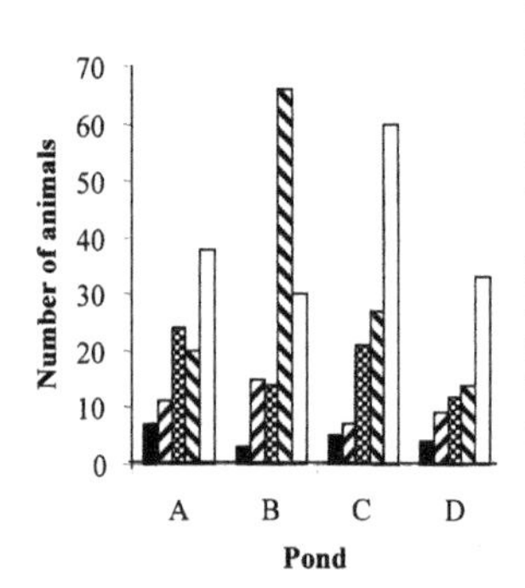

Fig. 111. Clustered bar graph of the numbers of animals in different insect orders found in ponds.

and figures for readers to examine the data in more detail if they wish.

More information about a range of other graphical techniques is given by Crothers (1981).

Writing up work

Much remains to be discovered about even the most basic aspects of the natural history of invertebrate animals and their interactions with their environment. Once a piece of research has been completed, it is important to communicate major results to others who are interested. Writing up is an important part of a research project. A thorough, critical investigation that has established new information of general interest may be worth publishing. Several journals publish short articles on investigations of this kind. They include the *Entomologist's Monthly Magazine* and various local publications such as *The Sorby Record,* which concentrates on work of relevance to Sheffield and the surrounding district. *The Journal of Biological Education* deals with articles of an educational nature. Examine a recent issue of the journal to see what sort of article they publish, and then write a paper on similar lines, keeping it as concise as possible while retaining enough information to establish the conclusions. Advice from an expert is helpful at this stage.

When you read papers written by other authors, bear in mind that if you do not understand something, others will not either. There is no merit in using jargon, and communication will be enhanced by keeping the work simple. Day (1998) describes in more detail how to write and publish scientific papers.

It is an unbreakable convention of scientific publication that results are reported with scrupulous honesty. It is therefore essential to keep detailed and accurate records throughout the investigation, and to distinguish in the write-up between certainty and probability, and between deduction and speculation.

Some useful addresses

Societies and publishers

The Amateur Entomologists' Society provides publications, news, and advice on the care and identification of insects and other invertebrates.
P.O. Box 8774, London SW7 5ZG
www.theaes.org

The Biological Records Centre is the national centre for species recording (excluding birds) and works with many other organisations on biodiversity mapping. It can also supply details of current local recorders and organisations involved in monitoring wildlife, such as the British Myriapod Group, the British Isopod Study Group and the British Arachnological Society.
NERC Centre for Ecology and Hydrology, Monks Wood, Abbots Ripton, Huntingdon PE28 2LS (01487 772400)
www.ceh.ac.uk/subsites/eic/brc

The British Entomological and Natural History Society supports amateur and professional entomologists and those interested in other branches of natural history, through meetings and publications including a journal (*The British Journal of Entomology and Natural History*).
The Pelham-Clinton Building, Dinton Pastures Country Park, Davis Street, Hurst, Reading, Berkshire RG10 0TH
www.benhs.org.uk

British Wildlife Publishing produces a journal (*British Wildlife*) and other conservation-orientated publications.
Lower Barn, Rooks Farm, Rotherwick, Hook, Hampshire RG27 9BG (01256 760663)
www.britishwildlife.com

English Nature is a government agency that promotes nature conservation throughout England through the publication of reports and documents and by issuing advice and permits and by protecting species and habitats by implementing various Acts of Parliament.
Northminster House, Peterborough PE1 1UA (01733 455000)
www.english-nature.org.uk

The Field Studies Council promotes environmental education and understanding through a network of field centres, and via courses and publications including the journal, *Field Studies*, and the AIDGAP series of identification guides.
Preston Montford, Montford Bridge, Shrewsbury SY4 1HW (01743 852100)
www.field-studies-council.org

The Freshwater Biological Association conducts research into freshwater science, houses a library, provides advice to members, and publishes a range of scientific publications including keys for the identification of freshwater organisms.
The Ferry House, Ambleside, Cumbria LA22 0LP
(01539 442468)
www.fba.org.uk

The Health and Safety Executive provides advice and support on health and safety, and a number of publications, including a brief introduction to COSHH called *COSHH: A Brief Guide to the Regulations*, which is available on-line free of charge.
Magdalen House, Trinity Road, Bootle, Merseyside L20 3QZ
(08701 545500)
www.hse.gov.uk

The Linnean Society of London covers all areas of biology, but especially diversity of, and the interrelationships between, organisms. The society houses a library and arranges scientific meetings, as well as publishing a number of journals and a series of identification guides, the *Synopses of the British Fauna*, that cover a wide range of invertebrate groups.
Burlington House, Piccadilly, London W1J 0BF
(02074 344479)
www.linnean.org

The Marine Biological Association of the United Kingdom promotes scientific research into marine life, arranges meetings and publishes a journal, *The Journal of the Marine Biological Association*.
The Laboratory, Citadel Hill, Plymouth, Devon PL1 2PB
(01752 633100)
www.mba.ac.uk

The Marine Conservation Society is concerned with the marine environment and its wildlife and acts as a pressure group as well as producing and distributing several publications on marine life.
9 Gloucester Road, Ross-on-Wye, Herefordshire HR9 5BU
(01989 566017)
www.mcsuk.mcmail.com

Pisces Conservation Ltd supplies some out-of-print books such as *Land and water bugs of the British Isles* by Southwood & Leston (1959) on CD-ROM.
IRC House, The Square, Pennington, Lymington, Hampshire SO41 8GN (01590 676622)
www.pisces-conservation.com

The Richmond Publishing Company Ltd supplies wildlife books including the *Naturalists' Handbooks* (which it also publishes), *AIDGAP* keys and the Linnean Society *Synopses of the British Fauna*.
P.O. Box 963, Slough SL2 3RS (01753 643104)
e-mail rpc@richmond.co.uk

The Royal Entomological Society disseminates information about insects and provides a communication network between entomologists through the publication of several journals (including *Ecological Entomology*, and *Agricultural and Forest Entomology*), a bulletin (*Antenna*) and a series of identification guides (*Handbooks for the Identification of British Insects*), as well as arranging meetings and maintaining a large library.
41 Queen's Gate, London SW7 5HR (02075 848361)
www.royensoc.demon.co.uk

The Wildlife Trust is a conservation charity coordinated by the Royal Society for Nature Conservation dedicated to conservation and has a network of local wildlife trusts and a junior branch (Wildlife Watch)
www.wildlifetrusts.org

Equipment suppliers

BioQuip Products supply equipment and books for entomology and related sciences.
17803 La Salle Avenue, Gardena, California 90248-3602, USA (310 324 0620)
www.bioquip.com

Edme Ltd. supply carbon dioxide dispensers.
Customer Services Department, Edme Ltd, Mistley, Manningtree, Essex CO11 1HG (01206 393725)
www.edme.com

GB Nets (Educational Field Equipment UK Ltd) supply field equipment, especially related to freshwater sampling.
P.O. Box 1, Bodmin, Cornwall PL31 1YJ (01752 295407)
www.gbnets-uk.com

Marris House Nets supplies nets, traps and other entomological equipment.
54 Richmond Park Avenue, Queens Park, Bournemouth BH8 9DR (01202 515238)
www.pwbelg.clara.net/marris

Philip Harris supplies field and laboratory equipment for catching invertebrates, and for viewing and storing them, as well as for monitoring other aspects of the environment.
Philip Harris Education, Customer Service Centre, Novara House, Excelsior Road, Ashby Park, Ashby De La Zouch, Leicestershire LE65 1NG (0870 6000193)
www.philipharris.co.uk

RS Components Ltd. supplies electronic components and equipment for monitoring the environment
P.O. Box 99, Corby, Northamptonshire NN17 9RS
(01536 201201)
rswww.com

The Tanglefoot Company supplies materials for catching insects and barriers to prevent invertebrate movement.
314 Straight Avenue, S. W., Grand Rapids, MI 49504-6485, USA (616 459 4139)
www.tanglefoot.com

Watkins and Doncaster supplies equipment for catching, setting and preserving insects
P.O. Box 5, Cranbrook, Kent TN18 5EZ (01580 753133)
www.watdon.com

Phil Wheater and Penny Cook have written a series of Microsoft Excel spreadsheet routines for calculating the statistics covered by this book. These illustrate the calculations using the data from the worked examples in this book and can be used to analyse your own data. These run in Excel 97-2000, and are available for £10 (including disk, instructions, spreadsheet routines, postage and packing) from the authors at:
The Department of Environmental and Geographical Sciences, The Manchester Metropolitan University, Manchester M1 5GD (0161 247 1589)
e-mail p.wheater@mmu.ac.uk

Appendix I
Calculating statistics

Calculating statistics by hand

When calculating statistics by hand, it is important to approach the calculations in the right order, and to know what each element of a formula means. As an example we take the fictitious formula below:

$$Q = \frac{|a - b|}{\sqrt{(c + d)} \times e^2}$$

If $a = 5$, $b = 8$, $c = 4$, $d = 2$, and $e = 3$, then:

$$Q = \frac{|5 - 8|}{\sqrt{(4 + 2)} \times 3^2} = \frac{|5 - 8|}{\sqrt{(4 + 2)} \times 9} = \frac{|5 - 8|}{\sqrt{(6)} \times 9} = \frac{|5 - 8|}{\sqrt{54}} = \frac{|5 - 8|}{7.3485} = \frac{|-3|}{7.3485}$$

$$= \frac{3}{7.3485} = 0.4082$$

The first thing to note is that any component of a formula in brackets, such as the addition of c and d at the bottom of the equation (in the denominator) is dealt with first. A squared number (e^2) is also treated as if it is in brackets, and calculated first. Components enclosed within a $|\ldots|$ symbol are dealt with in a similar way to components enclosed in brackets and the subtraction of $a - b$ on the top of the equation (the numerator) is completed before dealing with the $|\ldots|$ symbol which tells us to take the absolute value of anything within it (that is, to ignore any negative signs resulting from the calculation inside). When using a scientific calculator, the easiest way forward is to deal with the denominator first and calculate the square of 3 (= 9). This can be recorded in the memory of the calculator (using the 'M in' key) and the addition of 2 and 4 calculated next. Keeping the value of this (= 6) on the calculator screen, pressing 'x' and recalling the memory (using the 'MR' key) will multiply the 6 and 9 together (= 54). Pressing the '√' key will give the square root of 54 (= 7.3485), and this can then be recorded in the memory, overwriting the previous entry. The numerator can now be calculated by subtracting 8 from 5 (= -3) and removing the minus sign using the '±' key. The result (= 3) can be divided by the contents of the memory by pressing the '÷' key followed by the 'MR' key, giving the answer as 0.4082.

Another mathematical symbol that is commonly encountered in statistics is the $\sum$ symbol, which means take the sum of (that is, add together) all possible values. For example, $\sum x$ is the sum of all possible values of x. Note that $\sum x^2$ means the sum of all possible values of x^2, while $(\sum x)^2$ means all possible values of x are added together and the result is then squared. Again, the component within the brackets ($\sum x$) is completed before the result is used in another way (that is, the squaring calculation).

Using a computer program

A variety of computer programs ease the calculation of statistics. Unfortunately, very few enable all the statistics shown in this book to be calculated (many merely cover the tests in Chapter 5), and most are expensive. Programs commonly used in education and research establishments include SPSS and Minitab which share a common data entry format (for further details see Wheater & Cook, 2000 and Dytham, 2003). Many spreadsheets (for example Microsoft Excel) can be used to calculate these statistics, by inputting appropriate data and formulae. Some tests are also available either as part of the Excel spreadsheet itself, or as statistical add-ins to the spreadsheet. An easy-to-use Excel workbook is available from the authors (address, p. 100) that illustrates the calculations using the data from the worked examples in this book and can be used to analyse your own data.

Mathematical glossary

The following symbols appear in various formulae in this text. Box numbers are given for the first mention of each symbol. For each test or technique, we have used the symbols that are most frequently adopted by other publications and statistics packages. Sometimes this means that the same letter is used more than once. Where this happens the context is given and a box referred to.

$\sum R_1$ sum of the ranks in sample 1 (box 5.4)

$\sum x$ sum of all possible values of x (that is, all values of x added together) (box 4.2)

$\sum x^2$ sum of all possible values of x squared (that is, the values are squared before adding them together) (box 4.2)

$(\sum x)^2$ squared sum of all possible values of x (that is, the values are added together and then squared) (box 4.2)

$\sum xy$ sum of all possible values of x multiplied by the equivalent value of y (that is multiply each x value by its equivalent y value and then add them all up) (box 5.7)

d difference between each pair of data points in a paired t test (box 5.5)

d	absolute difference between ranks within each pair of data points in a Spearman's correlation coefficient (box 5.8)
d	density of organisms in the nearest neighbour method of examining distributions (box 4.6)
d	Berger–Parker index (box 4.7)
$\bar{d}$	mean difference between pairs of data (box 5.5) (known as '*d* bar')
D	Simpson diversity index (box 4.7)
df	degrees of freedom (see p. 67)
H'	Shannon diversity index (box 4.7)
J'	evenness index (box 4.7)
$\ln p_i$	the natural logarithm of the proportional abundance of each species at a site in a diversity index calculation (box 4.7)
$\ln S$	the natural logarithm of the number of species found at a site in a diversity index calculation (box 4.7)
M_1	number of animals captured, marked and released on day 1 in a Lincoln index population estimation (box 4.4)
M_2	number of animals captured, marked and released on day 2 in a Lincoln index population estimation (box 4.4)
n	sample size (number of data points) (box 4.2)
N	number of individuals at a site in a diversity index calculation (box 4.7)
N_{max}	number of individuals of the most abundant species in a dominance or diversity index calculation (box 4.7)
N_1	number of animals marked on day 1 and recaptured on day 2 in a Lincoln index population estimation (box 4.4)
p_{Ai}	percentage of species *i* in site A when calculating percentage similarities between sites (box 4.8)
p_{Bi}	percentage of species *i* in site B when calculating percentage similarities between sites (box 4.8)
p_i	proportional abundance of species *i* in a site in a diversity index calculation (box 4.7)
P	probability (p. 65)
$P_{A,B}$	percentage similarity between sites A and B (box 4.8)
$\hat{P}_1$	population estimate on day 1 in a Lincoln index population estimation (box 4.4)
$Q1$	quartile 1 (box 4.3)
$Q3$	quartile 3 (box 4.3)
r	test statistic for Pearson's correlation coefficient (box 5.7)
r_d	mean distance between nearest neighbours in the nearest neighbour method of examining distributions (box 4.6)

r_e	expected distance between nearest neighbours in the nearest neighbour method of examining distributions (box 4.6)
r_i	measured distance to the nearest neighbour from a particular individual in the nearest neighbour method of examining distributions (box 4.6)
r_s	test statistic for Spearman's correlation (box 5.8)
R	index of aggregation in the nearest neighbour method of examining distributions (box 4.6)
s	standard deviation of the sample (box 4.2)
S	total number of species found at a site in a diversity index calculation (box 4.7)
s^2	variance of the sample (box 4.2)
SE	Standard error (box 4.2)
$SE(\hat{P}_1)$	standard error of the population estimate in a Lincoln index population estimation (box 4.4)
t	test statistic for t test (box 5.1)
T_+	positive test statistic for Wilcoxon matched-pairs test (box 5.6)
T_-	negative test statistic for Wilcoxon matched-pairs test (box 5.6)
U	test statistic for Mann–Whitney U test (box 5.3)
x	data point (box 4.2)
$\bar{x}$	mean of x (box 4.2) (known as 'x bar')
χ^2	test statistic for chi square analysis (boxes 5.9 and 5.10)
y	data point (part of a pair of data points x, y) (box 5.7)
z	test statistic for z test (box 5.3)

Appendix II
Statistical tables

Table A1. *Random numbers (from 0 to 99)*. Select a starting point in the table at random and work along the row to find pairs of coordinates to position your samples. If the coordinates of your area do not run from 0 to 100, multiply or divide the numbers from the table by an appropriate factor. For example when surveying a 25 x 25 m area, divide each coordinate by 4.

1	64	43	2	8	3	11	98	73	13	40	97	76	99	64	77	96	68	18	46
40	96	87	57	77	46	5	46	14	71	22	99	83	12	74	31	94	92	44	99
68	83	34	56	98	94	99	81	29	12	60	24	3	76	81	60	43	55	78	48
24	53	56	19	17	69	44	50	85	48	20	24	26	96	30	34	50	98	42	4
64	48	42	5	11	65	80	32	79	30	74	45	97	84	21	54	87	2	12	78
3	85	94	7	61	70	9	80	56	66	63	83	60	46	24	16	60	81	16	84
98	75	70	45	44	38	15	24	9	65	97	72	10	45	23	76	20	76	73	61
58	8	59	10	84	33	22	92	33	18	37	85	81	97	56	84	25	63	57	72
18	92	51	99	84	77	96	24	9	75	99	34	46	50	22	59	25	94	69	54
55	15	49	15	8	36	2	44	27	81	40	50	28	62	36	49	55	68	63	14
74	81	18	11	53	45	50	37	76	45	79	38	62	32	59	41	32	51	68	72
5	92	2	26	76	79	2	48	11	90	52	86	20	61	61	78	67	66	26	73
17	99	19	43	71	84	85	74	7	60	61	46	82	4	4	99	95	24	20	33
85	17	81	67	89	18	66	39	54	39	64	62	95	86	82	46	60	49	67	74
83	52	22	73	65	81	62	25	13	62	36	15	78	15	90	5	34	97	21	68
35	84	68	64	38	69	23	96	81	6	64	31	15	88	12	76	99	41	85	26
71	78	37	67	75	30	7	91	23	7	30	56	72	45	45	87	62	96	43	93
52	99	95	33	96	47	61	54	19	92	97	32	83	34	68	98	20	71	27	26
74	88	35	57	31	39	88	72	94	24	53	96	54	81	95	89	91	97	35	13
97	33	69	12	9	75	78	97	89	53	41	81	97	75	87	11	45	75	89	82

Table A2. *Values of* t *and* z. If using these to calculate 95% confidence limits of the mean (box 4.2), there are $n - 1$ degrees of freedom. For an unpaired t test (box 5.1) or for a one-sample t test (box 5.2), there are $n_1 + n_2 - 2$ degrees of freedom (where n_1 and n_2 are the sizes of samples 1 and 2 respectively). For a z test (box 5.3) read the values at the bottom right hand side of the table where df = infinity (∞). For a paired t test (box 5.5), there are $n - 1$ degrees of freedom (where n is the number of data pairs). In t tests (paired or unpaired), the results are significant at the given probability level if the calculated value of t is higher than the table value.

df	*t*		df	*t*	
	***P* = 0.05**	*P* = 0.01		***P* = 0.05**	*P* = 0.01
1	**12.706**	63.657	17	**2.110**	2.898
2	**4.303**	9.925	18	**2.101**	2.878
3	**3.182**	5.841	19	**2.093**	2.861
4	**2.776**	4.604	20	**2.086**	2.845
5	**2.571**	4.032	21	**2.080**	2.831
6	**2.447**	3.707	22	**2.074**	2.819
7	**2.365**	3.499	23	**2.069**	2.807
8	**2.306**	3.355	24	**2.064**	2.797
9	**2.262**	3.250	25	**2.060**	2.787
10	**2.228**	3.169	26	**2.056**	2.779
11	**2.201**	3.106	27	**2.052**	2.771
12	**2.179**	3.055	28	**2.058**	2.763
13	**2.160**	3.012	29	**2.045**	2.756
14	**2.145**	2.977	30	**2.042**	2.750
15	**2.131**	2.947			
16	**2.120**	2.921	∞	**1.960**	2.576

Table A3. *Values of* U *for the Mann–Whitney* U *test*. The upper values in each row (in bold) are for $P = 0.05$, while the lower values are for $P = 0.01$. The results are significant at the given probability level if the lower calculated U value is equal to or lower than the table value.

n_S (number of items in the smaller data set)	n_L (number of items in the larger data set)																
	4	5	6	7	8	9	10	11	12	13	14	15	16	17	18	19	20
2	-	-	-	-	**0**	**0**	**0**	**0**	**1**	**1**	**1**	**1**	**1**	**2**	**2**	**2**	**2**
	-	-	-	-	-	-	-	-	-	-	-	-	-	-	-	0	0
3	-	**0**	**1**	**1**	**2**	**2**	**3**	**3**	**4**	**4**	**5**	**5**	**6**	**6**	**7**	**7**	**8**
	-	-	-	-	-	0	0	0	1	1	1	2	2	2	2	3	3
4	**0**	**1**	**2**	**3**	**4**	**4**	**5**	**6**	**7**	**8**	**9**	**10**	**11**	**11**	**12**	**13**	**14**
	-	-	0	0	1	1	2	2	3	3	4	5	5	6	6	7	8
5		**2**	**3**	**5**	**6**	**7**	**8**	**9**	**11**	**12**	**13**	**14**	**15**	**17**	**18**	**19**	**20**
		0	1	1	2	3	4	5	6	7	7	8	9	10	11	12	13
6			**5**	**6**	**8**	**10**	**11**	**13**	**14**	**16**	**17**	**19**	**21**	**22**	**24**	**25**	**27**
			2	3	4	5	6	7	9	10	11	12	13	15	16	17	18
7				**8**	**10**	**12**	**14**	**16**	**18**	**20**	**22**	**24**	**26**	**28**	**30**	**32**	**34**
				4	6	7	9	10	12	13	15	16	18	19	21	22	24
8					**13**	**15**	**17**	**19**	**22**	**24**	**26**	**29**	**31**	**34**	**36**	**38**	**41**
					7	9	11	13	15	17	18	20	22	24	26	28	30
9						**17**	**20**	**23**	**26**	**28**	**31**	**34**	**37**	**39**	**42**	**45**	**48**
						11	13	16	18	20	22	24	27	29	31	33	36
10							**23**	**26**	**29**	**33**	**36**	**39**	**42**	**45**	**48**	**52**	**55**
							16	18	21	24	26	29	31	34	37	39	42
11								**30**	**33**	**37**	**40**	**44**	**47**	**51**	**55**	**58**	**62**
								21	24	27	30	33	36	39	42	45	48
12									**37**	**41**	**45**	**49**	**53**	**57**	**61**	**65**	**69**
									27	31	34	37	41	44	47	51	54
13										**45**	**50**	**54**	**59**	**63**	**67**	**72**	**76**
										34	38	42	45	49	53	57	60
14											**55**	**59**	**64**	**69**	**74**	**78**	**83**
											42	46	50	54	58	63	67
15												**64**	**70**	**75**	**80**	**85**	**90**
												51	55	60	64	69	73
16													**75**	**81**	**86**	**92**	**98**
													60	65	70	74	79
17														**87**	**93**	**99**	**105**
														70	75	81	86
18															**99**	**106**	**112**
															81	87	92
19																**113**	**119**
																93	99
20																	**127**
																	105

Table A4. *Values of* T *for the Wilcoxon matched-pairs test.* The results are significant at the given probability level if the calculated value of T is less than or equal to the table value.

n (number of items minus any zero differences)	T		n (number of items minus any zero differences)	T	
	$P = 0.05$	$P = 0.01$		**$P = 0.05$**	$P = 0.01$
6	**0**	-	16	**29**	19
7	**2**	-	27	**34**	23
8	**3**	0	18	**40**	27
9	**5**	1	19	**46**	32
10	**8**	3	20	**52**	37
11	**10**	5	21	**58**	42
12	**13**	7	22	**65**	48
13	**17**	9	23	**73**	54
14	**21**	12	24	**81**	61
15	**25**	15	25	**89**	68

Table A5. *Values for Pearson's product moment correlation coefficient* (r). The results are significant at the given probability level if the calculated value of r is higher than the table value. n is the number of data pairs.

df	r		df	r	
$(n-2)$	**$P = 0.05$**	$P = 0.01$	$(n-2)$	**$P = 0.05$**	$P = 0.01$
1	**0.997**	1.000	16	**0.468**	0.590
2	**0.950**	0.990	17	**0.456**	0.575
3	**0.878**	0.959	18	**0.444**	0.561
4	**0.811**	0.917	19	**0.433**	0.549
5	**0.754**	0.874	20	**0.423**	0.537
6	**0.707**	0.834	21	**0.413**	0.526
7	**0.666**	0.798	22	**0.404**	0.515
8	**0.632**	0.765	23	**0.396**	0.505
9	**0.602**	0.735	24	**0.388**	0.496
10	**0.576**	0.708	25	**0.381**	0.487
11	**0.553**	0.684	26	**0.374**	0.479
12	**0.532**	0.661	27	**0.367**	0.471
13	**0.514**	0.641	28	**0.361**	0.463
14	**0.497**	0.623	29	**0.355**	0.456
15	**0.482**	0.606	30	**0.349**	0.449

Table A6. *Values for Spearman's rank correlation coefficient* (r_s). The results are significant at the given probability level if the calculated value of r_s is higher than the table value.

n	r_s		n	r_s	
	P = 0.05	P = 0.01		**P = 0.05**	P = 0.01
5	**1.000**	-	18	**0.472**	0.600
6	**0.886**	1.000	19	**0.460**	0.584
7	**0.786**	0.929	20	**0.447**	0.570
8	**0.738**	0.881	21	**0.436**	0.556
9	**0.700**	0.833	22	**0.425**	0.544
10	**0.648**	0.794	23	**0.416**	0.532
11	**0.618**	0.755	24	**0.407**	0.521
12	**0.587**	0.727	25	**0.398**	0.511
13	**0.560**	0.703	26	**0.390**	0.501
14	**0.538**	0.679	27	**0.383**	0.492
15	**0.521**	0.654	28	**0.375**	0.483
16	**0.503**	0.635	29	**0.368**	0.475
17	**0.488**	0.618	30	**0.362**	0.467

Table A7. *Values of Chi square* (χ^2). For a test of association (box 5.9) the degrees of freedom (df) are (the number of rows minus one) x (the number of columns minus one). For a goodness of fit test (box 5.10) the df are the number of categories minus one. The results are significant at the given probability level if the calculated value of χ^2 is higher than the table value.

df	χ^2		df	χ^2	
	P = 0.05	P = 0.01		**P = 0.05**	P = 0.01
1	**3.814**	6.635	16	**26.296**	32.000
2	**5.991**	9.210	17	**27.587**	33.409
3	**7.815**	11.345	18	**28.869**	34.805
4	**9.488**	13.277	19	**30.144**	36.191
5	**11.070**	15.086	20	**31.410**	37.566
6	**12.592**	16.812	21	**32.671**	38.932
7	**14.067**	18.475	22	**33.924**	40.289
8	**15.507**	20.090	23	**35.172**	41.638
9	**16.919**	21.666	24	**36.415**	42.980
10	**18.307**	23.209	25	**37.652**	44.314
11	**19.675**	24.725	26	**38.885**	45.642
12	**21.026**	26.217	27	**40.113**	46.963
13	**22.362**	27.688	28	**41.337**	48.278
14	**23.685**	29.141	29	**42.557**	49.588
15	**24.996**	30.578	30	**43.773**	50.892

References and further reading

Some of the books and journals listed here will not be available in local and school libraries. It is possible to make arrangements to see or borrow such works in several ways: by seeking permission to visit the library of a local university or of an appropriate society (such as the Royal Entomological Society), or by asking your local public library to borrow the work (or a photocopy of it) for you via the British Library's Document Supply Centre. This may take several weeks, and it is important to present the librarian with a reference that is correct in every detail. References are accepted in the form given here, namely the author's name and date of publication, followed by (for a book) the title and publisher or (for a journal article) the title of the article, the journal title, the volume number, and the first and last page numbers of the article.

Naturalists' Handbooks, AIDGAP keys and other Field Studies Council publications and Linnean Society Synopses of the British Fauna are available from The Richmond Publishing Co. Ltd, and Handbooks for the Identification of British Insects are available from the Royal Entomological Society (addresses, p. 99).

Allen, S. E. (1989, 2nd edn.) *Chemical analysis of ecological materials*. Oxford: Blackwell Scientific Publishers.

Armitage, P. D., Furse, M. T. & Wright, J. F. (1979) A bibliography of works for the identification of freshwater invertebrates in the British Isles. *FBA Occasional Publication No. 5*. Ambleside: Freshwater Biological Association.

Askew, R. R. (1988) *Identification chart of British and Irish dragonflies*. Colchester: Harley Books.

Barber, A. D. (1996) A key to the lithobiomorph centipedes of Britain. *Bulletin of the British Myriapod Group* **12**: 45–51.

Barber, A. D. & Keay, A. N. (1995) A tabular key to British geophilomorphs. *Bulletin of the British Myriapod Group* **11**: 30–31.

Barnard, P. C. (1999) *Identifying British insects and arachnids: an annotated bibliography of key works*. Cambridge: Cambridge University Press.

Barnes, R. S. K. (1994) *The brackish-water fauna of Northwestern Europe*. Cambridge: Cambridge University Press.

Bell, J. R., Wheater, C. P., Henderson, R. & Cullen, W. R. (2002) Testing the efficiency of suction samplers (G-vacs): the effect of increasing nozzle size and suction time. *Proceedings of the 19th European Colloquium of Arachnology, University of Aarhus, Denmark July 2000*.

Blackman, R. (1974) *Aphids*. London: Ginn and Company Ltd.

Blower, J. G. (1985) Millipedes. *Synopses of the British Fauna (New Series) No. 35*. London: The Linnean Society of London.

Blower, J. G., Cook, L. M. & Bishop, J. A. (1981) *Estimating the size of animal populations*. London: Allen & Unwin.

Brinkhurst, R. O. (1963) A guide for the identification of British aquatic Oligochaeta. *FBA Scientific Publication* 22. Ambleside: Freshwater Biological Association.

Brodie, J. (1985). *Grassland studies*. London: George Allen & Unwin Ltd.

Brooks, S. (1997) *Field guide to the dragonflies and damselflies of Great Britain and Ireland*. Hampshire: British Wildlife Publishing.

Brown, V. K. (1990, 2nd edn.) *Grasshoppers*. Naturalists' Handbooks 2. Slough: The Richmond Publishing Co. Ltd.

Bullock, J. (1996) Plants. Pp. 111–138 in: Sutherland W. J. (ed.) *Ecological census methods*. Cambridge: Cambridge University Press.

Burton, J. F. & Ragge, D. R. (1988) *Sound guide to the grasshoppers and allied insects of Great Britain and Ireland*. Colchester: Harley Books.

Cameron, R. A. D. & Redfern, M. (1976) British land snails. *Synopses of the British Fauna (New Series) No. 6*. London: The Linnean Society of London.

Cameron, R. A. D., Jackson, N. & Eversham, B. (1983) A field guide to the slugs of the British Isles. (FSC Publication 156.) *Field Studies*, **5**: 807–824. An AIDGAP Key.

Campbell, A. C. (1984) *The seashore and shallow seas of Britain and Europe*. London: Hamlyn.

Carter, D. J. & Hargreaves, B. (1986) *Caterpillars of Britain and Europe*. London: Collins.

Chinery, M. (1986) *Insects of Britain and Western Europe*. London: Collins.

Chinery, M. (1993, 3rd edn.) *A field guide to the insects of Britain and Northern Europe*. London: Collins.

Colyer, C. N. & Hammond, C. O. (1968, 2nd edn.) *Flies of the British Isles*. London: Frederick Warne and Co. Ltd.

Corbet, S. A., Fussell, M., Ake, R., Fraser, A., Gunson, C., Savage, A. & Smith, K. (1993) Temperature and the pollinating activity of social bees. *Ecological Entomology* **18**: 17–30.

Corke, D. (1999) Identifying insects. *British Wildlife* **10**(3): 153–163.

Courtecuisse, R. & Duhem, B. (1995) *Mushrooms and toadstools of Britain and Europe*. London: Collins.

Cranston, P. S. (1982) A key to the larvae of the British Orthocladiinae (Chironomidae). *FBA Scientific Publication 45*. Ambleside: Freshwater Biological Association.

Croft, P. S. (1986) A key to the major groups of British freshwater invertebrates. (FSC publication 181.) *Field Studies* **6**: 531–579. An AIDGAP Key.

Crothers, J. (1981) On the graphical presentation of quantitative data. *Field Studies* **5**: 487–511.

Crothers, J. (1997) A key to the major groups of British marine invertebrates. (FSC publication 246.) *Field Studies* **9**: 1–177. An AIDGAP Key.

Crothers, J. & Crothers, M. (1988, revised edn.) A key to the crabs and crab-like animals of inshore British waters. (FSC publication 155.) *Field Studies* **5**: 753–806. An AIDGAP Key.

Davis, B. N. K. (1991, 2nd edn.) *Insects on nettles*. Naturalists' Handbooks 1. Slough: The Richmond Publishing Co. Ltd.

Day, R. A. (1998, 5th edn.) *How to write and publish a scientific paper*. Cambridge: Cambridge University Press.

Disney, R. H. L. (1999) British Dixidae (meniscus midges) and Thaumaleidae (trickle midges): keys with ecological notes. *FBA Scientific Publication 56*. Ambleside: Freshwater Biological Association.

Dobson, F. S. (2000, 4th edn.) *Lichens*. Slough: The Richmond Publishing Co. Ltd.

Dobson, M. & Frid, C. (1998) *Ecology of aquatic systems*. London: Addison Wesley.

Du Feu, C., Hounsome, M. & Spence, I. (1983) A single-session mark/recapture method of population estimation. *Ringing and Migration* **4**: 211–226.

Dytham, C. (2003, 2nd edn.) *Choosing and using statistics: a biologist's guide*. Oxford: Blackwell Science.

Eason, E. H. (1964) *Centipedes of the British Isles*. London: Frederick Warne & Co. Ltd.

East, D. & Knight, D. (1998) Sampling soil earthworm populations using household detergent and mustard. *Journal of Biological Education* **32**: 201–206.

Eddington, J. M. & Hildrew, A. G. (1995) A revised key to the caseless caddis larvae of the British Isles, with notes on their ecology. *FBA Scientific Publication 53*. Ambleside: Freshwater Biological Association.

Edwards, C. A. & Bohlen, P. (1996, 3rd edn.) *Biology and ecology of earthworms*. New York: Kluwer Academic Publishing.

Edwards, C. A. & Lofty, J. R. (1977, 2nd edn.) *Biology of earthworms*. London: Chapman and Hall.

Elliott, J. M. (1977, 2nd edn.) Some methods for the statistical analysis of benthic invertebrates. *FBA Scientific Publication 25*. Ambleside: Freshwater Biological Association.

Elliott, J. M. (1996) British freshwater Megaloptera and Neuroptera. *FBA Scientific Publication 54*. Ambleside: Freshwater Biological Association.

Elliott, J. M. & Mann K. H. (1979) A key to the British freshwater leeches. *FBA Scientific Publication 40*. Ambleside: Freshwater Biological Association.

Ellis, A. E. (1978) British freshwater bivalve Mollusca. *Synopses of the British Fauna (New Series) No. 11*. London: The Linnean Society of London.

Emmet, A. M., Langmaid, J. & Heath, J. (1983 onwards) *The moths and butterflies of Great Britain and Ireland*. Colchester: Harley Books.

Erzinçlioğlu, Z. (1996) *Blowflies*. Naturalists' Handbooks 23. Slough: The Richmond Publishing Co. Ltd.

Fielding, A. H. & Howarth, P. F. (1999) *Upland habitats*. London: Routledge.

Fitter, R., Fitter, A. & Blamey, M. (1996, 5th edn.) *The wild flowers of Britain and Northern Europe*. London: Collins.

Fitter, R., Fitter, A. & Farrer, A. (1984) *Grasses, sedges, rushes and ferns of Britain and Northern Europe*. London: Collins.

Fitter, R. & Manuel, R. (1994) *Photoguide to lakes, rivers, streams and ponds of Britain and North West Europe*. Hong Kong: HarperCollins.

Fish, J. D. & Fish, S. (1996, 2nd edn.) *A student's guide to the seashore*. Cambridge: Cambridge University Press.

Forsythe, T. G. (2000, 2nd edn.) *Common ground beetles*. Naturalists' Handbooks 8. Slough: The Richmond Publishing Co. Ltd.

Friday, L. E. (1988) A key to the adults of British water beetles. (FSC publication 189.) *Field Studies* 7: 1–151. An AIDGAP Key.

Fry, R. & Waring, P. (1996) A guide to moth traps and their use. *Amateur Entomologist Volume 24*. Manningtree, Essex: Amateur Entomologist's Society Publications.

Gilbert, F. S. (1993, 2nd edn.) *Hoverflies*. Naturalists' Handbooks 5. Slough: The Richmond Publishing Co. Ltd.

Gledhill, T., Sutcliffe, D. W. & Williams W. D. (1993) British freshwater Crustacea: Malacostraca. *FBA Scientific Publication 52*. Ambleside: Freshwater Biological Association.

Gledhill, T. & Viets, K. O. (1976) A synonymic and bibliographic check-list of the freshwater mites (Hydrachnellae and Limnohalacaridae, Acari) recorded from Great Britain and Ireland. *FBA Occasional Publication No. 1*. Ambleside: Freshwater Biological Association.

Goater, B. (1986) *British pyralid moths: a guide to their identification*. Colchester: Harley Books.

Goddard, J. (2000) *Waterside guide*. Powys: Coch-y-Bonddu Books.

Graham, A. (1988, 2nd edn.) Molluscs: British prosobranch and other operculate gastropod molluscs. *Synopses of the British Fauna (New Series) No. 2*. London: The Linnean Society of London.

Greig-Smith, P. (1983, 3rd edn.) *Quantitative plant ecology*. Oxford: Blackwell Scientific Publications.

Guthrie, M. (1989) *Animals of the surface film*. Naturalists' Handbooks 12. Slough: The Richmond Publishing Co. Ltd.

Harde, K. W. (2000) *A field guide in colour to beetles*. Enderby: Silverdale Books.

Harker, J. (1989) *Mayflies*. Naturalists' Handbooks 13. Slough: The Richmond Publishing Co. Ltd.

Hayward, P. J. (1988) *Animals on seaweed*. Naturalists' Handbooks 9. Slough: The Richmond Publishing Co. Ltd.

Hayward, P. J. (1994) *Animals of sandy shores*. Naturalists' Handbooks 21. Slough: The Richmond Publishing Co. Ltd.

Hayward, P. J., Nelson Smith, A. & Shields, C. (1996) *Collins pocket guide. Sea shore of Britain and Europe*. London: HarperCollins.

Hayward, P. J. & Ryland, J. S. (eds.) (1990) *The Marine Fauna of the British Isles and North-West Europe*. Volumes 1 & 2. Oxford: Oxford University Press.

Hayward, P. J. & Ryland, J. S. (eds.) (1995) *Handbook of the Marine Fauna of North-West Europe*. Oxford: Oxford University Press.

Hellawell, J. M. (1986) *Biological indicators of freshwater pollution and environmental management*. London: Elsevier Applied Science Publishers.

Hillyard, P. D. (1996) Ticks. *Synopses of the British Fauna (New Series) No. 52*. London: The Linnean Society of London.

Hillyard, P. D. & Sankey, J. H. P. (1989, 2nd edn.) Harvestmen. *Synopses of the British Fauna (New Series) No. 4*. London: The Linnean Society of London.

Hingley, M. (1993) *Microscopic life in Sphagnum*. Naturalists' Handbooks 20. Slough: The Richmond Publishing Co. Ltd.

Hodge, P. & Jones, R. (1995) *New British beetles – species not in Joy's practical handbook*. Reading: British Entomological and Natural History Society.

Hopkin, S. (1991) A key to the woodlice of Britain and Ireland. (FSC publication 204.) *Field Studies* **7**: 599–650. An AIDGAP key.

Hopkin, S. (in press) A key to the springtails of Britain and Ireland. (FSC publication.) *Field Studies*. An AIDGAP key.

Hubbard, C. E. (1984, 3rd edn.) *Grasses*. London: Penguin Books.

Hynes, H. B. N. (1977, 3rd edn.) Adults and nymphs of British stoneflies (Plecoptera): a key. *FBA Scientific Publication 17*. Ambleside: Freshwater Biological Association.

Ingle, R. W. (1996, 2nd edn.) Shallow-water crabs. *Synopses of the British Fauna (New Series) No. 25*. London: The Linnean Society of London.

Jones-Walters, L. M. (1989) Keys to the families of British spiders. (FSC publication 197.) *Field Studies* **7**: 365–443. An AIDGAP key.

Joy, N. H. (1932) *A practical handbook of British beetles*. London: H. F. and G. Witherby. (E. W. Classey of Faringdon, Oxfordshire produced a reduced-sized reprint in 1997).

Keay, A. N. (1995) A dichotomous key to the geophilomorph centipedes of Britain. *Bulletin of the British Myriapod Group* **11**: 27–29.

Kennedy, J. & L. Crawley, L. (1967) Spaced-out gregariousness in sycamore aphids *Drepanosiphum platanoides* (Schrank) (Hemiptera, Callaphididae). *Journal of Animal Ecology* **36**: 147–163.

Kent, M. & Coker, P. (1992) *Vegetation description and analysis*. Chichester: John Wiley & Sons.

Kerney, M. P., & Cameron, R. A. D. (1979) *Land snails of Britain and North-West Europe*. London: Collins.

Killeen, I., Aldridge, D. & Oliver, G. (in press) *Freshwater bivalves of the British Isles.* (FSC publication.) An AIDGAP key.

Kirk, W. D. J. (1984) Ecologically selective coloured traps. *Ecological Entomology* **9**: 35–41.

Kirk, W. D. J. (1992) *Insects on cabbages and oilseed rape.* Naturalists' Handbooks 18. Slough: The Richmond Publishing Co. Ltd.

Kirk, W. D. J. (1996) *Thrips.* Naturalists' Handbooks 25. Slough: The Richmond Publishing Co. Ltd.

Krebs, C. J. (1999, 2nd edn.) *Ecological methodology.* California: Addison Wesley.

Leather, S. R. & Bland, K. P. (1999) *Insects on cherry trees.* Naturalists' Handbooks 27. Slough: The Richmond Publishing Co. Ltd.

Lee, D. G. & Corbet, S. A. (1989) Evaluating colonization samplers for freshwater invertebrates. *Journal of Biological Education* **23**: 23–31.

Legg, G. & Jones, R. E. (1988) Pseudoscorpiones (Arthropoda: Arachnida). *Synopses of the British Fauna (New Series) No. 40.* London: The Linnean Society of London.

Lincoln, R. J. (1979) *British gammaridean amphipods.* London: British Museum (Natural History).

Lincoln, R. J. & Sheals, J. G. (1979) *Invertebrate animals: collection and preservation.* London: British Museum (Natural History).

Lutz, F. E., Welch, P. L., Galtsoff, P. S. & Needham, J. G. (1959) *Culture methods for invertebrate animals.* New York: Dover Publications.

Macan, T. T. (1994, 4th edn., reprint.) A key to the British fresh- and brackish-water gastropods. *FBA Scientific Publication 13.* Ambleside: Freshwater Biological Association.

Magurran, A. E. (1988) *Ecological diversity and its measurement.* London: Croom Helm.

Majerus, M. & Kearns, P. (1989) *Ladybirds.* Naturalists' Handbooks 10. Slough: The Richmond Publishing Co. Ltd.

Manuel, R. L. (1983, 2nd edn.) *The Anthozoa of the British Isles – a colour guide.* Produced for the Marine Conservation Society by R. Earll. Ross-on-Wye: Marine Conservation Society.

Manuel, R. L. (1988, revised edn.) British Anthozoa. *Synopses of the British Fauna (New Series) No. 18.* London: The Linnean Society of London.

Marshall, J. A. & Haes, E. C. M. (1990, reprint with minor amendments.) *Grasshoppers and allied insects of Great Britain and Ireland.* Colchester: Harley Books.

Mason, C. F. (1996, 3rd edn.) *Biology of freshwater pollution.* Harlow: Longman.

McDaniel, B. (1979) *How to know the mites and ticks.* Dubusque, Iowa: Wm. C. Brown Company Publishers.

McLusky, D. S. (1990, 2nd edn.) *The estuarine environment.* Glasgow: Blackie.

Merryweather, J. & Hill, M. (1992, 2nd edn.) The fern guide. (FSC publication 211.) *Field Studies* **8**: 101–188. An AIDGAP Key.

Miller, P. L. (1995, 2nd edn.) *Dragonflies.* Naturalists' Handbooks 7. Slough: The Richmond Publishing Co. Ltd.

Mitchell, A. (1978) *Field guide to the trees of Britain and Northern Europe.* London: Collins.

Mitchell, A. & Wilkinson, J. (1991, 2nd edn.) *The trees of Britain and Northern Europe.* London: Collins.

Moore, P. D. & Chapman, S. B. (1986, 2nd edn.) *Methods in plant ecology.* Oxford: Blackwell Scientific Publications.

Morris, M. G. (1991) *Weevils.* Naturalists' Handbooks 16. Slough: The Richmond Publishing Co. Ltd.

Naylor, E. (1972) British marine isopods. *Synopses of the British Fauna (New Series) No. 3*. London: The Linnean Society of London.

Nelson-Smith, A. & Knight-Jones, P. (1990) Annelida: Polychaeta. In: Hayward, P. J. & Ryland, J. S. (eds.). *The Marine Fauna of the British Isles and North-West Europe*. Volume 1, Chapter 6. Oxford: Oxford University Press.

Nichols, D. (1999, 5th edn.) *Safety in biological fieldwork – guidance notes for codes of practice*. London: Institute of Biology.

Oliver, P. G. & Meechan, C. J. (1993) Woodlice. *Synopses of the British Fauna (New Series) No. 49*. London: The Linnean Society of London.

Olsen, L-H., Sunesen, J. & Pedersen, B. V. (2001) *Small freshwater creatures*. Oxford: Oxford University Press.

Phillips, R. (1980) *Grasses, ferns, mosses and lichens*. London: Macmillan.

Picton, B. E. (1993) *A field guide to the shallow-water echinoderms of the British Isles*. London: Immel Publishing.

Pigott, J. & Watts, S. (1996) Safety. In: S. Watts & L. Halliwell (eds.) (1996) *Essential environmental science*. London: Routledge.

Plant, C. (1997) A key to the adults of British lacewings and their allies. (FSC publication 245.) *Field Studies* **9**: 179–269. An AIDGAP key.

Porter, J. (1997) *The colour identification guide to the caterpillars of the British Isles (Macrolepidoptera)*. Middlesex: Viking.

Price, E. A. C. (2003) *Grassland habitats*. London: Routledge.

Prŷs-Jones, O. E. & Corbet, S. A. (1991, 2nd edn.) *Bumblebees*. Naturalists' Handbooks 6. Slough: The Richmond Publishing Co. Ltd.

Purvis, O. W., Coppins, B. J., Hawksworth, D. L., James, P. W. & Moore, D. M. (1998, 2nd edn.) *The lichen flora of Great Britain and Ireland*. London: The British Lichen Society.

Ragge, D. R. & Reynolds, W. J. (1998) *The songs of the grasshoppers and crickets of Western Europe*. Colchester: Harley Books.

Redfern, M. (1995, 2nd edn.) *Insects and thistles*. Naturalists' Handbooks 4. Slough: The Richmond Publishing Co. Ltd.

Redfern, M. & Askew R. R. (1992) *Plant galls*. Naturalists' Handbooks 17. Slough: The Richmond Publishing Co. Ltd.

Read, H. J. & Frater, M. (1999) *Woodland habitats*. London: Routledge.

Reynoldson, T. B. & Young, J. O. (2000) A key to the freshwater triclads of Britain and Ireland. *FBA Scientific Publication 58*. Ambleside: Freshwater Biological Association.

Richardson, D. H. S. (1992) *Pollution monitoring with lichens*. Naturalists' Handbooks 19. Slough: The Richmond Publishing Co. Ltd.

Roberts, M. J. (1993) *The spiders of Great Britain and Ireland*. Compact edition with Appendix of Addenda and Corrigenda. Parts 1 and 2. Colchester: Harley Books.

Roberts, M. J. (1995) *Field guide to the spiders of Britain and Northern Europe*. London: Collins.

Rose, F. (1981) *The wild flower key*. London: Frederick Warne.

Rotheray, G. E. (1989) *Aphid predators*. Naturalists' Handbooks 11. Slough: The Richmond Publishing Co. Ltd.

Salt, D. T. & Whittaker, J. B. (1998) *Insects on dock plants*. Naturalists' Handbooks 26. Slough: The Richmond Publishing Co. Ltd.

Savage, A. A. (1989) Adults of the British aquatic Hemiptera Heteroptera. *FBA Scientific Publication 50*. Ambleside: Freshwater Biological Association.

Savage, A. A. (1999) Keys to the larvae of British Corixidae. *FBA Scientific Publication 57*.

Ambleside: Freshwater Biological Association.

Scourfield, D. J. & Harding, J. P. (1966) A key to the British freshwater Cladocera. *FBA Scientific Publication 5*. Ambleside: Freshwater Biological Association.

Sims, R. W. & Gerard, B. M. (1999, 2nd edn.) Earthworms. *Synopses of the British Fauna (New Series) No. 31*. London: The Linnean Society of London.

Skinner, B. (1998, 2nd edn.) *Colour identification guide to moths of the British Isles (Macrolepidoptera)*. Middlesex: Viking.

Skinner, G. J. & Allen, G. W. (1996) *Ants*. Naturalists' Handbooks 24. Slough: The Richmond Publishing Co. Ltd.

Smalden, G. (1994, 2nd edn.) Coastal shrimps and prawns. *Synopses of the British Fauna (New Series) No. 15*. Revised and enlarged by L. B. Holthius & C. H. J. M. Fransen. London: The Linnean Society of London.

Smith, K.G.V. (1989) An introduction to the immature stages of British flies. *Handbooks for the Identification of British Insects*. London: Royal Entomological Society of London.

Snow, K. R. (1990) *Mosquitoes*. Naturalists' Handbooks 14. Slough: The Richmond Publishing Co. Ltd.

Soar, C. D. & Williamson, W. (1925, 1927, 1929) *The British Hydracarina*. Volumes 1, 2 and 3. London: The Ray Society.

Sokal, R. R. & Rohlf, F. J. (1995, 3rd edn.) *Biometry*. New York: W. H. Freeman & Co.

Southward, A. J. (1976) On the taxonomic status and distribution of *Chthamalus stellatus* (Cirripedia) in the north east Atlantic region: with a key to the common intertidal barnacles of Britain. *Journal of the Marine Biological Association of the United Kingdom* **56**: 1007–1028.

Southwood, T. R. E. & Henderson, P. A. (2000, 3rd edn.) *Ecological methods*. Oxford: Blackwell Science.

Southwood, T. R. E. & Leston, D. (1959) *Land and water bugs of the British isles*. London: Frederick Warne and Co. Ltd (out of print, but available on CD-ROM from Pisces Conservation Ltd, address p. 98).

Speight, M. C. D. (1986) Criteria for the selection of insects to be used as bio-indicators in nature conservation research. *Proceedings of the 3rd European Congress of Entomology, Amsterdam 1986*, 485–488.

Stace, C. (1997, 2nd edn.) *New flora of the British Isles*. Cambridge: Cambridge University Press.

Stednick, J. D. (1991) *Wildland water quality sampling and analysis*. California: Academic Press.

Stubbs, A. E. & Falk, S. (1996, 2nd edn. plus supplements 1 and 2.) *British hoverflies*. Reading: British Entomological and Natural History Society.

Tebble, N. (1976, 2nd edn.) *British bivalve seashells: a handbook for identification*. Edinburgh: HMSO.

Thompson, T. E. (1988, 2nd edn.) Molluscs: benthic opisthobranchs. *Synopses of the British Fauna (New Series) No. 8*. London: The Linnean Society of London.

Tilling, S. M. (1987) A key to the major groups of British terrestrial invertebrates. (FSC publication 187.) *Field Studies* **6**: 695–766. An AIDGAP Key.

Tolman, T. & Lewington, R. (1997) *Field guide to the butterflies of Britain and Europe*. London: Collins.

Trudgill, S. (1989) Soil types: a field identification guide. (FSC Publication 196.) *Field Studies* **7**: 337–363.

Unwin, D. M. (1978) Simple techniques for microclimate measurement. *Journal of Biological Education* **12**: 179–189.

Unwin, D. M. (1980) *Microclimate measurement for ecologists*. London: Academic Press.

Unwin, D. (1981) A key to the families of British Diptera. (FSC Publication 143.) *Field Studies* **5**: 513–553. An AIDGAP Key.

Unwin, D. M. (1988, revised edn.) A key to the families of British Coleoptera (beetles) and Strepsiptera. (FSC publication 166.) *Field Studies* **6**: 149–197. An AIDGAP Key.

Unwin, D. (2001) A key to the families of British bugs (Hemiptera). *Field Studies* **10**: 1–35. An AIDGAP Key.

Unwin, D. M. & Corbet, S. A. (1991) *Insects, plants and microclimate*. Naturalists' Handbooks 15. Slough: The Richmond Publishing Co. Ltd.

Wallace, I. D., Wallace, B. & Philipson, G. N. (1990) A key to the case-bearing caddis larvae of Britain and Ireland. *FBA Scientific Publication 51*. Ambleside: Freshwater Biological Association.

Watson, E. V. (1981, 3rd edn.) *British mosses and liverworts*. Cambridge: Cambridge University Press.

Wheater, C. P. (1999) *Urban habitats*. London: Routledge.

Wheater, C. P. & Cook P. A. (2000) *Using statistics to understand the environment*. London: Routledge.

Wheater, C. P. & Read H. J. (1996) *Animals under logs and stones*. Naturalists' Handbooks 22. Slough: The Richmond Publishing Co. Ltd.

Willmer, P. (1985) Bees, ants and wasps: a key to genera of the British Aculeates. (*FSC Occasional Publication 7*.) Preston Montford: Field Studies Council. An AIDGAP Key.

Wright, A. (1990) British sawflies (Hymenoptera: Symphyta): a key to adults of the genera occurring in Britain. (FSC Publication 203.) *Field Studies* **7**: 531–593. An AIDGAP Key.

Yeo, P. F. & Corbet S. A. (1995, 2nd edn.) *Solitary wasps*. Naturalists' Handbooks 3. Slough: The Richmond Publishing Co. Ltd.

Zahradník, J. (1998*a*) *A field guide in colour to insects*. Enderby: Blitz Editions.

Zahradník, J. (1998*b*) *A field guide in colour to bees and wasps*. Enderby: Blitz Editions.

Zar, J. H. (1999, 4th edn.) *Biostatistical analysis*. London: Prentice-Hall.

Index

CPSIA information can be obtained at www.ICGtesting.com
Printed in the USA
LVOW05s0826061015

457114LV00003B/3/P

9 781784 270827